AF443584

Database Systems in Science and Engineering

Database Systems in Science and Engineering

J R Rumble, Jr
National Institute of Standards and Technology
Gaithersburg, MD, USA

and

F J Smith
The Queen's University of Belfast
Northern Ireland, UK

Adam Hilger
Bristol, Philadelphia and New York

British Library Cataloguing in Publication Data

Rumble, John R.
    Database systems in science and engineering.
    1. Databases. Design & management
    I. Title II. Smith, F. J. (Francis James) *1950*
    005.74

ISBN 0-7503-0048-5

Library of Congress Cataloging-in-Publication Data

Rumble, J. R.
    Database systems in science and engineering / J.R. Rumble, Jr. and F.J. Smith.
    296p. 24cm.
    Includes bibliographical references and index.
    ISBN 0-7503-0048-5 : £30.00
    1. Data base management. 2. Science—Data bases. 3. Engineering—Data bases.
    I. Smith, F. J. II. Title.
    QA76.9.D3R86 1990
    502'.85574—dc20   90-37768
                                                                            CIP

Published under the Adam Hilger imprint by IOP Publishing Ltd
Techno House, Redcliffe Way, Bristol BS1 6NX, England
335 East 45th Street, New York, NY 10017-3483, USA
US Editorial Office: 1411 Walnut St, Suite 200, Philadelphia, PA 19102

Printed in Great Britain by Galliard (Printers) Ltd, Norfolk.

Dedicated to our families

Diane, Wendy, Jonathan

Ann, Owen, Rory, Michael, Brian, Una

Contents

Preface

The two authors worked together on a large scientific database system for the International Atomic Energy Agency in Vienna in 1979. The database software had been specially written for the project by a previous scientist because there was at that time no database management system which would have suited the very varied nature of the data being stored. In particular it was necessary to use variable length fields and variable length records throughout the database, and even today, 10 years later, it would not be easy to find a suitable database management system to store such data.

We were forced to use this special, once-off system which did most of what we wanted, but not everything that a modern database management system would have provided, and certainly not everything that we needed. We therefore had to spend a lot of our time making changes to the software to meet our requirements as they arose from day to day, mainly requirements concerned with the printing of the data in book form and making the data available for interrogation. About half of our time was spent writing software rather than spending time on our primary task, providing accurate and useful data to meet the needs of our users. Unfortunately the software was so specialized that our efforts had no application outside our own system.

Later we found that this was a universal problem in science and engineering; we learned this from our personal experience of several technical databases and from numerous reports of technical database systems at conferences, particularly at the CODATA conferences. An examination of almost any large organization employing scientists and engineers will show that many scientists and engineers will have built their own special data systems for their own data; and often you will find that adjacent groups, or even individuals within groups, will have written their own special software to handle their data, quite independently of others. Wheels will be reinvented many times over and the cost of the development of these separate systems will obviously be very high. Clearly resources are being wasted. On the other hand the accountants and managers of the company will be much more likely to be using the corporate database system for all of their data needs and will rarely be using special systems to suit individual data needs.

There are probably two reasons for this. The first is that all of the database systems on the market are designed and marketed for use primarily by people such as managers and accountants to help them handle finance, orders, invoices, stock control, and so on. The systems are therefore not suited to the scientist or engineer who wishes to store technical data, or even if they are suited, they are not marketed in a manner which makes it easy for the scientist or engineer to use them, i.e. the manuals are written in a form which does not make them easy to apply to the solution of technical problems.

The second reason that scientists and engineers so often use their own systems to handle their technical data is that they mistakenly think that this is easy to do. In contrast to the accountant or manager, they usually have a considerable knowledge of using and programming computers and believe that they have the ability to start to build their own system, even though as physicists or mechanical engineers or biologists or chemists, they have no training in the building of a database system. When they decide to build their own system they will also find, if they look at the books on database systems in their library, that they are all written about the kind of databases that their accountant or management use. The way that these books are written and the examples chosen do not relate easily to the problem of storing, retrieving and manipulating technical data. Since we believe that there are no textbooks on the market which address this problem we agreed to pool our experiences to write this book. Our aim, however, was not primarily to help scientists or engineers to write their own new database systems, but rather to encourage them to use existing database management systems and to do so effectively; in the process we explain how these database management systems work.

If a database system is to be effective then it has to provide information to some group of users. It may be that the group of users consists only of the person building the database, but usually it is for a much wider group of people. In either case it is self-evident that if a database is to be successful then it must meet some need of the people who are going to use it. We have therefore put a large emphasis on the users of the database system and consequently on both the uses and nature of technical data and of technical databases. We have devoted a whole chapter to the user interface and another to the dissemination of data to users. We have emphasized in later chapters the necessity to find out what the users need when a database is being planned and designed.

We have included a chapter on expert systems as they are often associated with databases in science and engineering. A short section on object-oriented databases puts this new development in context for the non-specialist reader.

In summary, therefore, we have written this book in the hope that it will be a help to the scientist or engineer who wants to build a technical database system. It explains database technology in terms of technical data rather than commercial data, with examples chosen from science and engineering. It emphasizes that if the users are not satisfied, no database system can be successful; so careful systematic planning is a necessary prerequisite of a successful system.

J R Rumble, Jr

F J Smith

Acknowledgments

The authors would like to thank their many colleagues who have made numerous suggestions and comments. In particular, we would like to acknowledge Jean Gallagher, Marguerite Lennon, William Ruff, Ed Begley, Dave Anderson, Leslie Emerson, Johnny Tripathy and John Flanigan for their careful reading of the manuscript and for their constructive comments, which have greatly strengthened the final text. Our thanks are also due to Brian Smith who drew most of the figures using MACDRAW. Also Jeanne Bride, Mary Trapane, Diane Finlay and the secretarial center at Queen's University deserve thanks for their assistance with the preparation of the manuscript.

We also acknowledge J G Kaufman of the MPD Network, Columbus, Ohio, and the Standard Reference Data Program of the National Institute of Standards and Technology for permission to use screens from their database systems.

One of us, FJS, would like to thank the University of Connecticut, because a substantial part of his contribution to the book was written while on a visiting Professorship in the Department of Computer Science, University of Connecticut. In particular, Fred Maryanski, Head of Department, deserves thanks for his useful and encouraging comments at that time.

Professor Peter Gray of Aberdeen is due our thanks for supplying two relevant papers at short notice at the proof stage.

Chapter 1

Introduction to Technical Databases

1.1 INTRODUCTION

Almost every scientist and engineer now has access to a computer terminal or personal computer, and they naturally would like these to be a means of storing, manipulating and accessing technical data. Even though powerful software tools exist that make this possible, rarely have the tools been developed with technical data in mind. Rarer still are books and articles which provide guidance.

This book is aimed at filling this void by presenting to a technically trained person the whys and wherefores of database management in science and engineering. Because of the abundance of books that discuss the details of database management from the perspective of business and nontechnical data, our discussion will use scientific and engineering examples to illustrate important concepts. We will concentrate on the unique features of technical data, on the building and usage of technical databases, and on practical approaches to common problems.

1.2 TECHNICAL DATABASE MANAGEMENT

The subject of this book is the handling of scientific and engineering data on computers. For brevity, we will use throughout the book the word 'technical' as a synonym for 'scientific and engineering'. The common name for this activity is 'technical database management'.

1.2.1 Definition of database

Database has been defined by Date [1] as

> a collection of related data stored on a computer that can be used for different applications without knowledge of storage details.

This definition contains several key elements.

The data must be **related**. A collection of chemical names combined with daily stock market statistics would not be a set of related data, while chemical names and their molecular weights would be.

The data must be **computerized**. The word database is also used for published collections of data. We will reserve the term database for computerized collections.

Different applications can use databases without knowledge of the **storage details**. Data are retrieved by reference to the data or to data names rather than by resorting to format specification as done in FORTRAN and other programming languages. In essence, a database frees users from the limitations and detailed specification requirements associated with **data files**. Thus, users are able to access the needed information without resorting to low-level programming languages. Instead, higher-level languages and the data names themselves are used. Figure 1.1 illustrates the difference.

Unfortunately, in practice the term database is used with considerable latitude and the variations are numerous [2]. Examples include published data collections and self-contained programs that calculate data from a set of input parameters. Finally, other terms, databank, datapack, etc, are used in place of database, often in an attempt to distinguish between types of databases. The use of these words varies from country to country and in different situations. In this book we will use the term **database** to conform to Date's definition but recognize that in technical database practice, other types of data collections also use this name. Many of the planning and design techniques discussed here can also be applied to these collections.

In Fortran ....

```
 20   READ (1,100) NUMBER, NAME, AMW
100   FORMAT (5X,I5,5X,A10,5X,F6.4)
      IF (NAME.NE.'ACETYLENE') THEN GOTO 20
 30   CONTINUE
```

Using a database system ...

```
Find: mol wgt for name = acetylene
```

Figure 1.1 The molecular weight of acetylene from a database can be found without knowing that the molecular weight is the third field in a record stored as a floating point number with five significant figures and up to four decimal places.

1.2.2 *Database management systems*

The software system that manages a database is called a **database management system** (DBMS). The functions of a DBMS are discussed in more detail later, particularly in Chapters 8 and 9, but they control all operations on the database such as:

<table>
<tr><td>★ storage</td><td>★ input</td></tr>
<tr><td>★ retrieval</td><td>★ editing</td></tr>
<tr><td>★ searching</td><td>★ updating</td></tr>
<tr><td>★ display</td><td>★ correction</td></tr>
<tr><td>★ user interface</td><td>★ output to other programs</td></tr>
</table>

This is the main tool for using a database and involves techniques and concepts quite different from usual computer programming. In the last 10 years, many database management systems have been built and brought onto the market, especially relational database management systems (see Section 9.1). At the present time, a well-designed DBMS exists for almost all types of computers. Rarely will a database builder have to seriously consider writing an entirely new package. Scientific and engineering database management does require significant specialized facilities, but even these can be approached by adding to an existing system rather than developing something entirely new.

It is a primary thesis of this book that with proper understanding of the database needs and of existing database systems combined with good planning, databases for scientific and engineering data can be built with available tools.

1.2.3 Personal computing

The last 10 years have also seen a revolution in the computing resources that scientists and engineers now have on their desks. In the late 1970s, a few lucky users had dumb terminals that provided interactive access to some remote computer. Scheduling of jobs was totally out of the user's control, and often layers of people and procedures prevented full exploitation of computer power. This has been changed by the widespread availability of personal computers (PCs), and the latest models now provide more power sitting on a desk top than was available at some universities just 20 years ago. Speed and CPU size are very affordable and mass storage continues to drop in price. Through local area networks (LANs) and telecommunications, PCs can also access every level of computer resource.

One result of this is the opportunity for individual scientists and engineers to build and control data collections for their own needs and purposes. The many benefits are obvious. Therefore, one feature of this book is the discussion of technical database management for the PC as well as for the minicomputer and mainframe.

1.3 WHY DISCUSS TECHNICAL DATABASE MANAGEMENT?

The increased availability of personal computers is not enough to account for the rapidly increasing interest in technical databases. A most significant driving force is the computerization of every aspect of science and engineering. Equipment is completely computerized and often data are generated automatically in computerized form. Computer-aided design, analysis and manufacture are now routine, and finally, computer networks linking all these tools together have been developed.

Although storing data on a computer provides greater efficiency and compatibility throughout the course of technical work, the computerization of technical data has lagged considerably behind the other technical work mentioned above. The technology of technical database management has improved tremendously in the last decade, but building technical databases is not yet a trivial or routine matter.

Technical databases are significantly different from other databases. Technical data are more complex with greater structure and have severe representational problems. Chapter 3 discusses these matters in full detail, and it suffices here to point out that, for most technical areas, the details of handling data in databases are not yet agreed upon. Just as important, the use of technical data is not well understood. In most cases, technical data are used in the context of solving problems for which data themselves are only one part of the solution. Very little work has been done to develop a model of *uses* of technical data, especially in engineering.

The combinations of these factors make it timely to examine technical database management in detail. This book points out the common understanding and conventional wisdom as well as problems and pitfalls; it then suggests future avenues of research. The common features of technical data, from biology to chemistry, from civil engineering to mechanical engineering, far outweigh the obvious differences. The techniques for handling data from one technical area can be adapted to another technical area much more easily than the techniques from business or finance can be adapted to science or engineering.

1.4 PREVIOUS WORK ON TECHNICAL DATABASE MANAGEMENT

There have been few previous studies of technical database management. Rumble and Hampel [2] edited the only book-length study on the subject. Several other authors discussed many of the important issues: Shoshoni and colleagues [3] did important work on classifying technical databases; McCarthy [4] studied technical metadata and the concept of a data thesaurus; Smith, Hughes and collaborators [5] investigated the details of technical database management for several specific types of data; and Tubbs [6] and Hilsenrath [7] developed general database management systems in the early 1970s specifically for technical data.

In contrast to general techniques, there is an abundance of literature on specific technical database projects. Probably the most studied issue is the representation of chemical nomenclature and structure [8]. The American Chemical Society has sponsored numerous symposia on this subject as well as on chemical databases in general. Materials databases have also been thoroughly studied and a complete bibliography has been given by Westbrook [9]. Kaufman wrote a guide for materials databases [10]. The Committee on Data for Science and Technology of the

International Council of Scientific Unions (CODATA) holds biennial conferences on technical data in all disciplines, and the proceedings [11] of these conferences include many papers on technical databases.

1.5 THE USERS OF TECHNICAL DATABASE MANAGEMENT

A wide variety of people are concerned currently with technical database management. Publicly available technical databases usually involve a collaboration of computer scientists and technical specialists. Small databases built by individuals may involve only one person and a manual. For any technical database project, the problems and concerns are similar, though the scale can differ greatly. The methodology given in this book to address these concerns is also the same in principle regardless of the size.

We have written this book to appeal to anyone involved in technical database management. Scientists and engineers should profit because important computer science concepts are presented using examples drawn from technical areas. Computer scientists will gain because the peculiarities of technical data are shown along with their consequence to the databases. We hope users will come away with an understanding of the inner workings of technical databases as well as with increased expectations for their ease of use and performance and their contributions to the efficiency and accuracy of technical projects.

1.6 THE ORGANIZATION OF THIS BOOK

This book is broadly divided into two sections. The first section (Chapters 2–6) is intended to give the reader an overview of technical database activity and discusses why these databases are built and how they are used. Understanding this will allow readers to fit their particular database projects into perspective and to estimate the amount of effort needed. This pre-planning is crucial in keeping technical database projects under control and in paving the way for a successful effort. The dissemination of technical databases, an issue not treated in most database books, is discussed here also.

The second section (Chapters 7–12) addresses the technology of technical database management. Most examples of the concepts are drawn from technical data, usually from actual database projects. This is in contrast to traditional database books where the examples are not relevant to the work of scientists and engineers.

The new concepts presented in this book are left to a minimum, but are sufficient to provide the basis both for using database management systems on personal computers and for sorting through the inevitable jargon used to describe many database management systems.

Finally, we recognize that at every stage of the database development, users should be consulted because their use of a database is the final test. Users may be wrong but they are the consumers, and consideration of their opinions is essential for success.

REFERENCES

[1]   C. J. Date, *An Introduction to Database Systems*, 4th Edition, Addison-Wesley Publishing Co., Reading, MA, 1987.

[2]   J. R. Rumble, Jr. and V. E. Hampel, *Database Management in Science and Technology*, North Holland, Amsterdam, 1984.

[3]   A. Shoshani, F. Olken and H. Wong, Data Management Perspective of Scientific Data, in *The Role of Data in Scientific Progress, Proceed. 9th CODATA Conference*, Ed. P. G. Glaeser, North-Holland, Amsterdam, 1985.

[4]   J. L. McCarthy, Information Systems Design for Materials Properties Data, in *Computerization and Networking of Materials Databases*, Eds. J. S. Glazman and J. R. Rumble, Jr., American Society for Testing and Materials, Philadelphia, PA, 1989.
J. L. McCarthy, The Automated Data Thesaurus: A New Tool for Scientific Information, in *Proceedings of the Eleventh International CODATA Conference*, Ed. P. S. Glaeser, Hemisphere Press, New York, 1990.

[5]   F. J. Smith and J. G. Hughes, Some special features of a materials database system, in *Proceedings of DKSEM Conference on Data and Knowledge Systems for Manufacturing and Engineering*, Harford CT, Oct. 19-20, 1987, Computer Society of the IEEE, Washington DC 1988.

[6]   *Generalized Data Management Systems and Scientific Information*, Ed. N. Tubbs, OCED Nuclear Energy Agency, Paris, France, 1978.

[7]   J. Hilsenrath and B. Breen, OMNIDATA, An Interactive System for Data Retrieval, Statistical and Graphical Analysis, and Database Management, A User's Manual, *NBS Handbook 125*, U. S. Department of Commerce, Washington DC, 1978.

[8]   *Graphics for Chemical Structure*, Ed. W. Warr, *American Chemical Society Symposium Series No. 341*, American Chemical Society, Washington DC, 1987.

[9]   H. Wawrousek, J. H. Westbrook and W. Grattidge, Data sources of mechanical and physical properties of engineering material, *Physik Daten – Physics Data Series No 130-1*, Fachinformationzentrum Energie, Physik, Matematik Gmbh, Karlsruhe, FRG, 1989.

[10] Guide to material property database management, Ed. J. G. Kaufman, *CODATA Bulletin, No 69*, CODATA, Paris, France, Nov. 1988.

[11] *Proceedings of the Eleventh International CODATA Conference*, Ed. P. S. Glaeser, Hemisphere Press, New York, 1990.

Computer Handling and Dissemination of Data, *Proceedings of the Tenth International CODATA Conference*, Ed. P. S. Glaeser, Elsevier Science Publishing Co, Amsterdam, 1987.

The Role of Data in Scientific Progress, *Proceedings of the Ninth International CODATA Conference*, Ed. P. S. Glaeser, North-Holland, Amsterdam, 1985.

Data for Science and Technology, *Proceedings of the Eighth International CODATA Conference*, Ed. P. S. Glaeser, North-Holland, Amsterdam, 1983.

Data for Science and Technology, *Proceedings of the Seventh International CODATA Conference*, Ed. P. S. Glaeser, North-Holland, Amsterdam, 1981.

Proceedings of the Sixth International CODATA Conference, Ed. B. Dreyfus, Pergamon Press, Oxford, 1979.

Proceedings of the Fifth International CODATA Conference, Ed. B. Dreyfus, Pergamon Press, Oxford, 1977.

Generation, Compilation, Evaluation, and Dissemination of Data for Science and Technology, *Proceedings of the Fourth International CODATA Conference*, Ed. B. Dreyfus, Pergamon Press, Oxford, 1975.

Chapter 2

The Nature of Technical Data

2.1 BASIC CONSIDERATIONS

You have only to pick up a piece of scientific or engineering literature to realize that significant differences exist between it and everyday publications such as the daily newspaper. Common words have specialized meanings, additional characters (Greek and mathematical) are used, numbers are different, and schematic images are prevalent. In this chapter, we will outline the characteristics of scientific and engineering data and describe classification schemes for them. To illustrate these concepts, examples from the field of tribology (the study of wear and friction) will be used.

The purpose of this discussion is to give builders of technical databases insights into the special problems that are posed in handling technical data. The considerations include how the data are stored, what kinds of search strategies are needed, and what kinds of output displays are useful. Because technical data contain more data types than are normally handled in business and other nontechnical databases, technical database builders must often come up with solutions not covered by conventional database management systems.

2.2 TYPES OF TECHNICAL DATA

The characteristics of technical databases derive from both the nature of technical data themselves and the way in which technical data are generated. Technical data consist of seven primary types as shown in Table 2.1.

Almost any technical publication routinely contains the first six and often the last of these data types. The major problem in technical database management is storing and presenting all these data types in a manner which is technically sound and appealing to the user. The term 'factual' is often used, especially in Europe, to describe databases containing these technical data types.

In representing the knowledge, different subjects place emphasis on different data types. For example, biology puts more emphasis on text and pictures and less

8

Table 2.1 Primary types of technical data.

Text	Names, words and records using the 26 Latin letters plus punctuation and text layout
Numbers	Real, integer, error estimates, very large, very small, complex, imaginary
Scientific text	Names, words, records and mathematics using Greek and other characters, superscripts and subscripts of all characters
Relationships	Equations, correlations
Rules	Interpretation and use, constraints, methods of application
Representational images	Chemical structure diagrams, graphics, diagrams, biological drawings
Pictures	Photographic images, half tones

on relationships and numbers than does physics. In addition, the data types are often heavily intermingled. Frequently, the same data are described in two or more ways such as numbers, equations and graphs.

2.3 EXAMPLES FROM TRIBOLOGY

Tribology is the science and engineering of relative motion between two bodies in contact and includes information on such familiar subjects as friction, wear, gears and bearings. Tribological testing usually follows a specific test procedure which is often laid down by a standards organization. Variations in these test procedures are common. The tests involve a defined specimen of the material of interest moving in a prescribed manner over a second surface. The contact geometry between two material surfaces is defined, and sometimes additional substances, either lubricants or abrasive particles, are introduced to improve movement (lubricants) or simulate service conditions (abrasives). The different technical data types will be illustrated using examples from tribology data sources.

Figures 2.1(a) and 2.1(b) display sets of typical wear data for several materials commonly used in bearings and gears. Examples of the first three data types are shown, including text, numbers and scientific text. Note the use of both superscripts and subscripts as well as Greek letters.

Representational images are important in reporting tribology data. For example, Figure 2.2 shows a typical graph of the variation of a property (weight loss due to wear) as a function of an independent variable (carbon content). Such graphs are universally found in science and engineering. Figure 2.3 is the phase diagram for an alloy of tribological interest in electrical contacts. Unlike the simple x–y plot in Figure 2.2, many different curves covering different ranges are shown. These are difficult to read. In addition, the regions under the curves have significance that must be retained when this graph is computerized.

Class	Sub-class	Common Name	Grade	Form	Process Treatment	UNS	Hard-ness	Scale	Density kg/m3	Young's Modulus MPa	Wear Coef	Wear Rate	Wear Const mm2/N	Load N	Velocity m/s	Distance m	Test Number	Abrasive Flow Rate g/m	Abrasive Size
Metal	Ferrous	Steel	17-4PH	bar	HT	S17400	44	RC	7750	196480	160×10^{-5}	167×10^{-4}	372×10^{-9}	45	2.4	4309		435	AFS 50/70
Metal	Ferrous	Steel	17-4PH	bar	HT	S17400	44	RC	7750	196480	282×10^{-5}	851×10^{-4}	654×10^{-9}	130	2.4	1436	G65-B	329	AFS 50/70
Metal	Ferrous	Steel	316	bar	Annealed	S31600	97	RB	8027	196480	42.0×10^{-5}	86.7×10^{-4}	195×10^{-9}	45	2.4	4279	G65-D	234	AFS 50/70
Metal	Ferrous	Steel	316	bar	Annealed	S31600	97	RB	8027	196480	90.5×10^{-5}	187×10^{-4}	420×10^{-9}	45	2.4	4252		471	AFS 50/70
Metal	Nonferrous	Stellite	S1016	bar	Weld	—	52	RC			14.6×10^{-5}	33.6×10^{-4}	270×10^{-9}	45	2.4	4358		197	AFS 50/70
Metal	Nonferrous	Stellite	S1016	bar	Weld	—	52	RC			19.0×10^{-5}	43.8×10^{-4}	35.2×10^{-9}	124	2.4	4253		471	AFS 50/70
Metal	Nonferrous	Stellite	S1016	bar	Weld	—	52	RC			15.8×10^{-5}	38.1×10^{-4}	29.3×10^{-9}	124	2.4	4309	G65-A	239	AFS 50/70
Composite	Metal-Matrix	Tungsten Carbide	K-714	plate	Sintered	—	85	RA			3.22×10^{-5}	4.54×10^{-4}	3.65×10^{-9}	130	2.4	4252		471	AFS 50/70

Figure 2.1(*a*) A typical set of tribological abrasive wear data. (Thanks to W. Ruff).

Material	Treatment/Coating	Vickers Hardness kg/mm²	Relative Wear*
Carbon Steel	Annealed	180	165
Ni–Cr	Annealed	275	124
Cr–Steel	Hardened-Tempered (500 °C)	300	111
Cr–Steel	Hardened-Tempered (150 °C)	840	72
Carbon Steel	Bath Nitrided	640	73
Carbon Steel	WC-Co Plasma Coated	800–1000	106
Carbon Steel	Al_2O_3 Plasma Coated	850–950	127
Carbon Steel	Cr_2O_3	750–900	159
Carbon Steel	Dull Ni Plate (50 m)	560	106
Carbon Steel	Hard Cr Plate (50 m)	1050	23

* Relative wear in mg/4 hr.

Figure 2.1(*b*) Typical abrasive wear data for steel.

METALLURGY OF FERROUS MATERIALS FOR WEAR APPLICATIONS

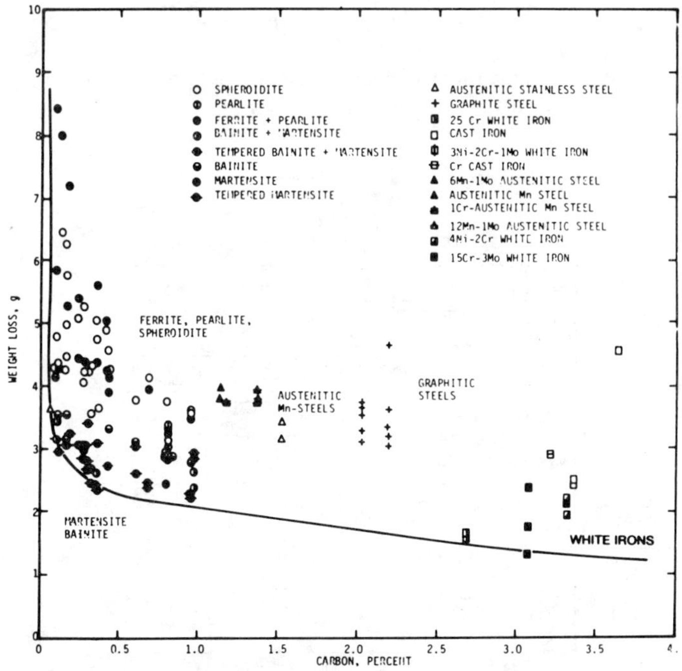

WEIGHT LOSS IN A WET SAND SLURRY ABRASION TEST OF A VARIETY OF FERROUS MATERIALS PLOTTED AGAINST THE CARBON CONTENT [16]

Figure 2.2 A typical *x–y* graph of technical data.

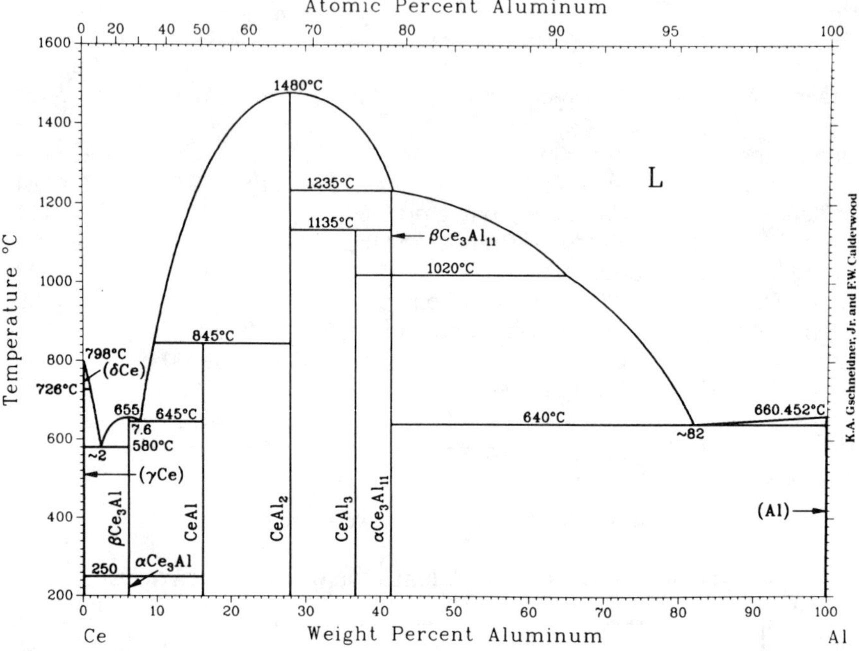

Figure 2.3 Another type of graph for technical data. In this plot, the areas under the curves have meaning as well as the curves and points themselves.

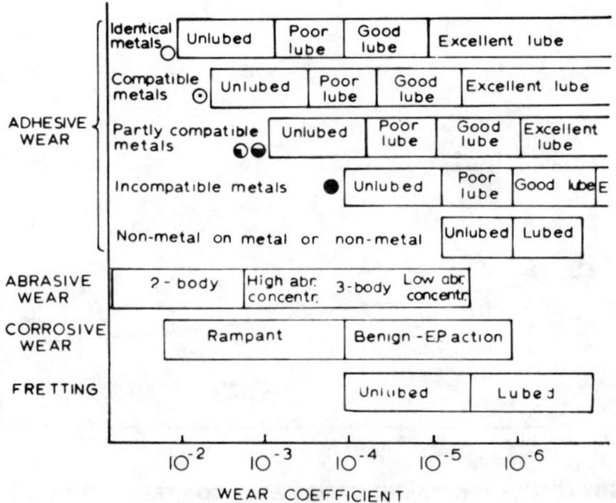

Figure 2.4 A diagrammatic plot of technical data. Here quantitative data are presented in a very qualitative manner.

A more diagrammatic example is shown in Figure 2.4 in which quantitative data are displayed in a qualitative way. Finally, another typical diagram, the chemical structure of a typical lubricant component, is shown in Figure 2.5.

Relationships found in technical data are exemplified in Figure 2.6. Pictures of tribological surfaces, Figure 2.7, are typical of those found in technical data sources. Finally, rules often accompany actual data, as shown in the example in Figure 2.8, which concerns the selection of lubricants.

These examples clearly indicate the complexity of technical data and the challenges that await the technical database builder with respect to entering the data into a computer and storing them such that all the original attributes of the data can be recovered. In the next section, we discuss how special features of technical data affect access to them in databases.

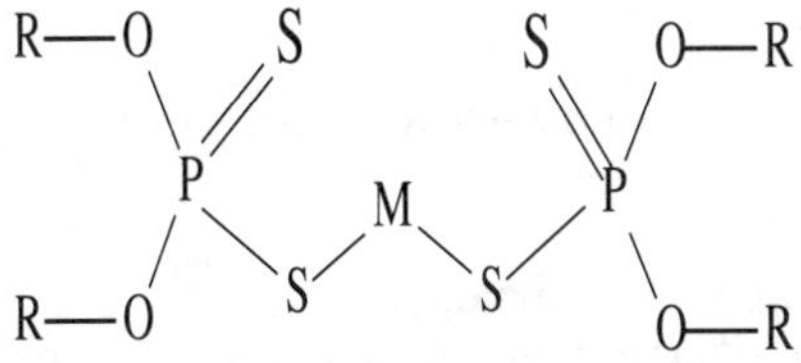

Figure 2.5 The chemical structure of metal dialkyldithiophosphates, common lubricant additives used in tribological application. The structure diagram is of great importance in chemistry. M is usually zinc but may also be molybdenum, tungsten, or other metals. The R–O– groups are derived from primary and secondary alcohols and alkyl phenols and may be single or mixed.

2.4 FEATURES OF TECHNICAL DATA THAT IMPACT ACCESS

Many features make scientific and engineering data difficult to access and retrieve in database management situations. These are summarized in Table 2.2. Examples will be drawn from Figures 2.1 to 2.8.

1. Size of numbers

Numbers can be of any type and range from very small to very large. For example, wear constants in conventional units are small, ranging from 10^{-7} to 10^{-8} (Figure 2.1(a)). Databases must be able to store, search and display the full range of numerical values. Scientific notation should be displayed in a clear way.

2. Interpolation among independent variables

Property data may vary, with one or more independent variables often measured at non-uniform intervals, sometimes widely dispersed. Most data requests are for values of the independent variables not present. In Figure 2.1(a) abrasive flow rates

Wear Coefficients for Phosphonates in Paraffinic Oil

Test Rotation Speed 1500 rpm
Test Load 15 kg
Temperature 50 C Time 1 hr

Additive Concentration = 4 mmol/100 g in Paraffinic Oil

Phosphonates: $(R_1O)_2P(O)R_2$

R_1:R_2	WSD (mm)	K ($¥$ 10^{-8})
No Additive	0.72	22.9
Ethyl:H	0.70	20.4
Butyl:H	0.64	14.1
Lauryl:H	0.32	0.64
Butyl:Hexyl	0.59	10.1
Butyl:Phenyl	0.48	4.2

The equation to calculate the wear volume of the rotating 12.7 diameter ball is:

$$V = (15.5\ ¥\ 10^{-6}D^3 - 1.03\ ¥\ 10^{-8}L)D, \text{cm}^3$$

where D is the wear scar diameter (WSD) in mm and L is the test load in kg. The wear coefficient K is calculated from the relation

$$K = (VH)/(2.33\ ¥\ (\text{rpm})\ ¥\ (t)\ ¥\ (0.408L))$$

where t is the time in minutes and H is the hardness, here taken as 725 kg/mm^2.

Figure 2.6 A typical relationship for technical data. The wear coefficient K is calculated from values of H = hardness, t = time and V = wear volume. The wear volume is itself calculated from D = wear scar diameter (WSD) and L = test load. Results for phosphonate lubricants are shown in the accompanying table.

were set at 234, 471 and other non-uniform values. Most data requests would be for more uniformly spaced flow rates, 300, 350, etc. When the requested independent variable values are missing, reliable schemes for interpolation must be used to satisfy the request.

3. Qualitative independent variables

Although independent variables are sometimes given qualitatively or imprecisely, there is often a need to search for precise values. The treatment coating information in Figure 2.1(*b*) is given in qualitative terms. The list of terms would be difficult to search unless all possible values were displayed.

4. Non-integer data values and range searching

Dependent variable values are real numbers, not integers, with a specific number of significant figures. It is usually not possible to search for them exactly. The wear constants (Figure 2.1(*a*)) for 316 stainless steel under different test conditions are 195×10^{-9} and 420×10^{-9} mm^{2}N^{-1}. However, when searching for wear constants,

Figure 2.7 A photograph of the surface of a specimen tested for abrasive wear data. Measurements are made on the size and shapes of the grooves that are eventually translated into qualitative data (with thanks to W. Ruff).

Rules for Lubricant Selection (Partial List)

The best lubricant for a particular application will be the simplest and least expensive which will meet the requirements.

In general the simplest and least expensive lubrication is a small quantity of plain mineral oil placed in the assembly without any feed system.

For choice of mineral oils, see section ...

A small quantity of mineral oil will not work when the life required is too long or wear debris accumulates too greatly or too much heat is generated. A change should be made to an oil feed system., see section ...

Where cooling is not a problem, but sealing is, the lubricant should probably be a grease, see section ...

Where the temperature range is too great for a mineral oil, or where a special requirement exists, such as low flammability, the best choice may be a synthetic oil, see section ...

Figure 2.8 A portion of rules governing the selection of a lubricant. The sections indicated in the answers to each question contain more detailed rules to guide the final selection. Such rules are readily amenable to expert systems.

Table 2.2 Features of technical data that impact access.

1. Size of numbers
2. Interpolation amongst independent variables
3. 'Qualitative' independent variables
4. Non-integral data values and range searching
5. Relationships between independent and dependent variables
6. Extrapolation
7. Qualitative representation of data
8. Precision
9. Complicated variable names and data values
10. Substance description
11. Substance equivalency
12. Units

users are likely to request data values such as 200×10^{-9} to 300×10^{-9} $mm^2\,N^{-1}$. Databases need to support searching over ranges of possible values so that requested data values can be found.

5. Relationship between independent and dependent variables

Relationships between dependent and independent variables are often expressed as precise mathematical relations or as fitted mathematical functions. In these cases, database queries become a matter of calculation rather than of search. For example, in Figure 2.6, wear coefficient values should be calculated using the equations rather than looked up in the table.

6. Extrapolation

Data may be required for a wide range of values of independent variables that fall outside of values contained in the database. If a legitimate method for doing this exists, such as an extrapolation formula based on a physical law, then this should be included in the database.

7. Qualitative representation of data

Technical data may be presented semiquantitatively as diagrams, schematics, pictures, or descriptive text that are difficult to quantify and whose parts must be searchable. Figure 2.4 is an excellent example. The terms 'poor lube' and 'good lube' are qualitative. The problem is whether or not this information can be quantified and put into a database in a form suitable for searching.

8. Precision

The range of precision and accuracy of technical data varies widely from precise measurements to rough estimates. Many physical measurements can be made to six or more significant figures. In contrast, corrosion rate data are so variable that they are usually reported as a range, e.g., 2 to 20 mils per year. For both precise and imprecise data, databases must maintain the correct number of significant figures throughout all operations including display. Imprecise data often have associated error ranges which extend the data values to the limits of those ranges and complicate searching and display.

9. Complex variable names and data values

Variable names often contain Greek letters and superscripts and subscripts that complicate searching, retrieval and display. Almost all the examples of technical data have such complications.

10. Substance description

Property data are associated with a particular material or substance. Most materials and substances can be described in several different ways by means of different designation systems, and to many different levels of detail. Substance descriptions are areas of intense interest in technical database management. Chemical structure and nomenclature have been extensively studied and many systems have been implemented [1]. Research on systems for biological materials and engineering materials is not as advanced but is equally important.

Materials and chemicals can be described and classified on several hierarchical levels as illustrated by the data in Figure 2.1(a). Data sets 1 and 2 should be found if any of the following queries are asked:

  (i) wear coefficients for stainless steel 316
 (ii) wear coefficients for all steels
(iii) wear coefficients for all ferrous alloys
(iv) wear coefficients for all metals

because stainless steel 316 is a steel, a type of ferrous alloy which is a metal. Good technical databases should allow for searching of materials at all levels and with different nomenclature systems if widely used.

11. Substance equivalency

Users often want to combine data from different data sources related to the same substance. To do this, they must be sure that the substances reported in the different sources are equivalent. The problem is compounded because the materials may be described using different nomenclature systems and may include varying levels of detail. For example, in Figure 2.1(a), many data fields are needed to determine if the substance on line 1 is identical to that on line 2.

12. Units

Numbers normally have units, and many different units are often possible for the same quantity. Units conversion can be carried out by the DBMS, thus facilitating search, display and comparison. However, retrieval in the units of the original measurement is often desirable and significant figures must be maintained throughout conversion.

While data types from other disciplines also encompass many of these features, we feel that only technical data contain them all on a routine basis. For a given database, these special features must be taken into account.

2.5   THE CLASSIFICATION OF TECHNICAL DATA

The characteristics of technical databases were studied by Shoshani, Olken and Wong [2], and a classification scheme as shown in Table 2.3 was devised that is useful in understanding different classes of technical data. Their work primarily looked at technical data from the point of view of the experimentalist or tester.

In the scheme of Shoshani *et al*, scientific and engineering data result from experimental measurements, from standardized test procedures, from calculations (also called simulations or modelling) and from theory. For simplicity, all such data are called **experimental data**. In almost every case, these experimental data are enhanced in significant ways. First, **support** data are needed that precisely define the conditions under which the data were measured (calculated, tested, etc). Then, the experimental results are processed further, resulting in more refined and usable information called **generated** data.

Alternatively, these data types can be looked at from the viewpoint of the user who in most cases needs **property** data to help in some particular application or problem-solving exercise. Property data are either intrinsic properties of a material, i.e., density, molecular weight, or characteristics resulting from a standardized test procedure, e.g., wear constants from standard tribology tests. Property data are thus obtained as the culmination of a long process that began with individual experimental measurements. In this chapter we will examine the data classes from the experimentalist's point of view, while in Chapter 3 we will adopt that of the user.

To illustrate the different data classes, we will use a common tribology test procedure in which the rate of material removal from the surface of a sliding metal block is measured. In the test, a block slides at a constant velocity over another material called the counterface. For our example this will be a hard plastic. Sand particles are added between the blocks to simulate real performance. The test is run several times under different conditions; the three test parameters that are changed are the test temperature, the size of the sand particles and the distance the block slides. The amount of material removed can be measured as a mass loss and is related to the sliding distance, temperature and size of the sand particles. From these data, the wear rate can be calculated. For this example, the dependent variable is the mass loss. The primary independent variables are test temperature, distance and sand particle size. Of course, there are many other independent variables, such as

Table 2.3 Classification of technical data.

Experiment or Test data

Support data

Configuration data

Instrumentation data

Generated data

Analyzed data

Summary or Handbook data

Property or Evaluated data

the pressure of one surface on the other, but only the above-named three change during different evaluations in this test set. Let us use this example to illustrate the concepts outlined in Table 2.3.

2.5.1 Support data

Support data are of two types: configuration data or instrumentation data.

Configuration data describe an experiment or simulation as it exists at the beginning of a measurement. They usually do not change over time, but when they do, these changes must be recorded. **Instrumentation** data contain information on the instruments and materials used in an experiment and how they change as it progresses.

For our tribological experiment, the configuration data included items such as the area of the block face in contact with the counterface, the velocity and the load. Naturally, other types of information could be included if they were deemed sufficiently relevant. For example, the test being used is Standard Test A1234. This calls for a block of 16 square inches (4 by 4), but the test was actually run with a block of 100 square centimeters (10 by 10). Then the test standard number (A1234) and the test variation (10 by 10 cm rather than 4 by 4 in) would be part of the configuration data. These data stay constant throughout the test.

Instrumentation data, in theory, contain a complete record of all conditions and changes through the course of a measurement, but, in actual fact, many readings are monitored sporadically. For the test described above, the three measurements that might be made throughout the test would be surface temperature, weight loss and distance travelled. These become the primary independent variables of the data for this experiment.

Every technical experiment, test and calculation contain both configuration and instrument data. Because the configuration data remain constant throughout the experiment, their size is small, while instrument data can be very large, especially if measurements are made frequently. Both types of information must be reported, and as the results from many different experiments are gathered together, analyzed, summarized and transformed to property data, the completeness of the data becomes even more important.

2.5.2 Generated data

Generated data result from the refinement of experimental data. Naturally, all the configuration and instrument data from each set of experimental data underlie each of the categories of generated data, but often the details are lost, obscured, or not noted.

Analyzed data come from the process of analyzing all relevant measurements for a given experiment and deriving the relationships between various dependent and independent variables. Depending on the scientific or engineering field involved, the type of analysis commonly employed differs significantly. Sometimes the analysis just involves a correction procedure for experimental data. In cases where the phenomena are well understood and scientific laws are known, the analysis might involve a straightforward transformation of experimental data. In other cases, statistical analysis is required or empirical analysis is made to determine the dependence of one variable on another.

In our example, suppose the mass loss was determined by cleaning the surface with a solvent and cleaning cloth and that the actual mass at any time was the measured mass minus a correction for a small amount of mass lost during cleaning. Then the actual mass loss would be the analyzed quantity. The wear rate data would be determined by analysis of the actual mass loss as a function of temperature and sand particle size. In this case, some standard equation might be used or the data simply put through a least-squares analysis. It should be noted that in practice, both configuration data and instrumentation data are necessary for this analysis.

Summary data come from the combination of many experimental data sets. For our tribology data, if we had repeated this measurement with several different materials and perhaps even several different kinds of sand and had made many tests for each combination, we could have produced a summary similar to Table 2.4, where typical wear rate data are given for common materials.

For both analyzed data and summary data, it is important to document the procedures and the original experimental data. However, the original experimental data are often not available. Probably the experimental data behind Table 2.4 could not be found. But presumably, the data were derived from statistical and other analysis of many different measurements.

Databases allow us the opportunity to give the user access to all classes of data, and for several database systems now under development, original experimental data and their analyzed counterparts, as well as the summary data, are made

Table 2.4 Sliding wear rates for common materials.

Type	Material	Sliding Wear Rate†	Temperature Limitation
Cast irons	Ni–hard martensitic white iron	0.100	no
	High phosphorus pig iron	0.323	no
Ceramics	Fusion cast alumina–zirconia–silica	0.0531	no
	Plate glass	0.81	no
Rubber-like plastics	Polyurethane	2.27–5.35	yes

† in units of sq in of material worn per 1000 tons of coke per sq ft of contact area.

available. Of course, caution must be taken to make sure that inappropriate comparisons between the two kinds of data are not made.

Property or evaluated data result from the collection of information measured by many groups over the years. They are usually organized according to materials and substances. Because of the wide range of coverage, these sources have become the usual first choice for users, even though the data have gone through several stages of analysis and summary and may not be easily traceable to a given set of measurements.

Property data for our tribological measurements would result from a project to analyze critically all sliding wear rate data for stainless steels. The result would be design values issued by an authoritative body, possibly as a handbook.

2.6   OTHER CLASSIFICATION SCHEMES FOR TECHNICAL DATA

In some cases, it is useful to classify technical data further in terms of their regularity, density and time variation (Shoshoni [2]).

Regularity means that measurements have been made according to some pattern. This implies that there can be a relationship between the measured data values and storage locations for them. An example would be determining the wear rate at intervals of ten degrees in temperature. Then only the initial temperature and the interval need be stored. For experiments where the number of measurements is very high, a major savings in data storage needs can be achieved. Often measurements are made at irregular intervals. Then complete support data are needed for each data value.

Density, or sparse identification, refers to those cases where there are large numbers of independent variables and many are unknown, or measured infrequently. This feature is a consideration when data are combined from more than one source and many data points have very sparse coverage. If sparseness exists, storage can be greatly reduced.

Time is a common independent variable and many technical data are time series with numerous data points only differing in their time of measurement. When storing this information, the relationship of the data points as a function of time becomes an important consideration. If the data points are mostly independent of time, then data compression can be applied effectively. Time series, though important for technical data, are not as common as they are in areas such as economics or census statistics.

2.7   TECHNICAL DATA AND METADATA

Technical information in databases, just as in other types of information, is a combination of data and metadata. Data are represented as data items which consist of the item name and a data value. Metadata are the information describing data items such as attribute names and units. The concept has been introduced to technical data by McCarthy [3] and Smith [4].

Examples of technical data have already been given, and in Chapter 7 the structure of data will be discussed further. Here we will use Figure 2.1(*a*) to illustrate the concept of metadata. Metadata essentially are all the information necessary to define data items, entities, and relationships within a specific database. In Figure 2.1(*a*), certain elements can be identified directly as metadata, namely the definitions of all column headings such as the terms 'class,' 'common,' 'grade' and 'Young's modulus' and the units 'MPa'. A typical metadata entry is shown in Figure 2.9 which includes not only a definition of Young's modulus but other related information. Figure 2.10 shows a similar entry for the units 'MPa.'

```
Term                = Tensile Modulus
Standard_Term       = Tensile Modulus
Type                = Property
Description         = The ratio of stress to corresponding strain
                      below the proportional limit
Abbreviation        = tnmod
Symbol              = E
Broader_Term        = Tensile Properties
Used_For            = tension modulus of elasticity
                    = tensile modulus
                    = Young's modulus
                    = static modulus of elasticity
                    = modulus of elasticity
Standard_Units      = pascal
```

Figure 2.9 Definition of Young's Modulus as provided by the online system of the MPD Network.

```
Term                = Pressure Units
Standard_Term       = Pressure Units
Type                = Units
Description         = The units in which pressure, strength and
                        stress are reported. The standard unit is
                        the pascal which is 1 joule/m³
Used_For            = pressure units
                    = strength units
                    = stress units
Standard_Units      = pascal
Abbreviation        = Pa
Valid_Units         = J/m³
                    = ksi
                    = psi
                    = MPa
                    ...
```

Figure 2.10 The definition of the unit pascal, also from the MPD Network.

The distinction between data and metadata is not always clear. A database might have a data thesaurus that provides explanatory information about data values themselves, which might primarily be text. For example, in Figure 2.1(*b*), the term **carbon steel** might be defined in the data thesaurus as shown in Figure 2.11. In other situations, the data values might be best explained by means of a set of additional data items. In Figure 2.1(*a*), the entry for 17-4PH steel could be described either by means of metadata (Figure 2.12) or by additional data fields (Figure 2.13).

Metadata can be searched and manipulated in a similar fashion to data themselves; for example, units can be changed. When individual databases are combined together into database systems or networks, metadata handling capability becomes important and needs special attention in a separate thesaurus.

2.8 DATA THESAURUS

As McCarthy [3] has pointed out, the concepts of data and metadata do not go far enough if several databases are considered at one time, as in the case of database systems or networks. In these cases, the data item names and metadata from individual databases are organized and combined to support search and retrieval on a system-wide basis. A data thesaurus is needed because multiple nomenclature and terminology systems occur in every technical discipline. Many different units can be used. Substance names can have many synonyms. Even property names are often not unique.

Compare, for example, the data in Tables 2.5 and 2.6. Both are legitimate representations of mechanical property data for materials of tribological interest

2.2 Carbon Steels

2.2.0 COMMENTS ON CARBON STEELS

2.2.0.1 *Metallurgical Considerations.*--Carbon steels are those steels containing carbon up to about 1 percent and only residual quantities of other elements except those added for deoxidation.

The strength that carbon steels are capable of achieving is determined by carbon content and, to a much lesser extent, by the content of the residual elements. Through cold working or proper choice of heat treatments, these steels can be made to exhibit a wide range of strength properties.

The finish conditions most generally specified for carbon steels include hot-rolled, cold-rolled, cold-drawn, normalized, annealed, spheroidized, stress-relieved, and quenched-and-tempered. In addition, the low-carbon grades (up to 0.25 percent C) may be carburized to obtain high surface hardness and wear resistance with a tough core. Likewise, the higher carbon grades are amenable to selective flame hardening to obtain desired combinations of properties.

2.2.0.2 *Manufacturing Considerations*

Forging.--All of the carbon steels exhibit excellent forgeability in the austenitic state provided the proper forging temperatures are used. As the carbon content is increased, the maximum forging temperature is decreased. At high temperatures, these steels are soft and ductile and exhibit little or no tendency to work harden. The resulfurized grades (free-machining steels) exhibit a tendency to rupture when deformed in certain high-temperature ranges. Close control of forging temperatures is required.

Cold Forming.--The very low-carbon grades have excellent cold-forming characteristics when in the annealed or normalized conditions. Medium-carbon grades show progressively poorer formability with higher carbon content, and more frequent annealing is required. The high-carbon grades require special softening treatments for cold forming. Many carbon steels are embrittled by warm working or prolonged exposure in the temperature range from 300 to 700 F.

Machining.--The low-carbon grades (0.30 percent C and less) are soft and gummy in the annealed condition and are preferably machined in the cold-worked or the normalized condition. Medium-carbon (0.30 to 0.50 percent C) grades are best machined in the annealed condition, and high-carbon grades (0.50 to 0.90 percent C) in the spheroidized condition. Finish machining must often be done in the fully heat-treated condition for dimensional accuracy. The resulfurized grades are well known for their good machinability. Nearly all carbon steels are now available with 0.15 to 0.35 percent lead, added to improve machinability. However, resulfurized and leaded steels are not generally recommended for highly stressed aircraft and missile parts because of a drastic reduction in transverse properties.

Welding.--The low-carbon grades are readily welded or brazed by all techniques. The medium-carbon grades are also readily weldable but may require preheating and postwelding heat treatment. The high-carbon grades are difficult to weld. Preheating and postwelding heat treatment are usually mandatory for the latter, and special care must be taken to avoid overheating. Furnace brazing has been used successfully with all grades.

2.2.0.3 *Environmental Considerations.*--Carbon steels have poor oxidation resistance above about 900 to 1,000 F, and protective atmospheres must be employed during heat treatment if scaling of the surface cannot be tolerated. Also, these steels are subject to decarburization at elevated temperatures and, where surface carbon content is critical, should be heated in reducing atmospheres. Strength and oxidation-resistance criteria generally preclude the use of carbon steels above 900 F.

Carbon steels exhibit a fairly abrupt drop in notch toughness as the service or testing temperature is lowered below room temperature. The subject of the notch toughness of steels is reviewed in references 2.2.0.3(a) and (b).

The corrosion resistance of carbon steels is relatively poor; clean surfaces rust rapidly in moist atmospheres. Simple oil film protection is adequate for normal handling. For aerospace applications, the carbon steels are usually plated to provide adequate corrosion protection.

Figure 2.11 The definition of the term carbon steel from a well-known handbook.

and, in fact, both have exactly the same data for precisely the same material. But all the names, units and data values are different. A search of a database system for elastic modulus data should find both entries even if they were in two separate databases. The function of the data thesaurus is to facilitate such retrieval by recognizing that the following equivalencies exist:

elastic modulus = modulus of elasticity
ksi and megapascals are units of the same dimensions
17-4PH stainless = S17400.

The most useful data systems allow the metadata to be managed in the same way as the data themselves. They can be searched, definitions displayed and equivalencies found. The power of the data thesaurus will grow as more databases are made available in systems.

```
Standard_Term          = S17400

Type                   = material

Short_Description      = a martensitic (semi-austenitic) precipi-
                         tation hardenable stainless steel of mod-
                         erate strength

Description            = This alloy is a martensitic stainless
                         steel, which can be precipitation hard-
                         ened to a wide range of mechanical
                         properties by a simple aging treatment in
                         the range of 900 F to 1150 F. Along with
                         high strength, it offers a good combina-
                         tion of corrosion resistance, fabricat-
                         ing characteristics and toughness. Like
                         other martensitic steels, it undergoes a
                         ductile-brittle transition at sub-zero
                         temperatures. Some typical applications
                         are aircraft and missile fittings, fas-
                         teners, gears, jet engines parts, valve
                         parts, chemical process equipment, pump
                         shafts, nuclear reactor components, and
                         paper mill equipment.

Broader_Term           = ph stainless steels
                       = age hardening steels
                       = stainless steels

Used_For               = 17-4PH
                       = ASTM A564
                       = ASTM A693
                       = S17400
                       = Cb-7 Cu-1
```

Figure 2.12 The definition of 17-4PH stainless steel in the MPD Network thesaurus.

```
Unified Number        S17400

Description           Precipitation Hardenable, Cr-Ni-Cu stain-
                      less Steel (17-4 PH)

Chemical
Composition           C 0.07 max, Cb 0.15-0.45, Cr 15.50-17.50,
                      Cu 3.00-5.00, Mn 1.00 max, Ni 3.00-5.00, P
                      0.040 max, S 0.030 max, Si 1.00 max

Cross Reference
  Specifications      AISI S17400, AMS 5604; 5622; 5643; 5825,
                      ASME SA564 (630); SA705 (630), ASTM A564
                      (630), A705 (630), MIL SPEC MIL-C-24111;
                      MIL-S-81506; MIL-S-81519, SAE J467 (17-
                      4PH)
```

```
UNS = S17400, max C = 0.07, min C = 0.00, max Cb =0.45, min Cb
= 0.15, max Cr = 17.50, min Cr = 15.50, max Cu = 5.00, min Cu
= 3.00, max Mn = 1.00, min Mn = 0.00, max Ni = 5.00, min Ni =
3.00, max P = 0.040, min P = 0.000, max S = 0.030, min S = 0.000,
max Si = 1.00, min Si = 0.00, AISI spec = S17400, AMS spec =
5604, AMS spec = 5622, AMS spec = 5643, AISI spec = 5825, ASME
spec = SA564, ...
```

Figure 2.13 An alternative definition of the steel 17-4PH as a series of data fields. The top of the figure shows the entry from the definition of this material; the bottom of the figure shows this as a series of data fields. Compare this entry to the metadata entry in Figure 2.12.

Table 2.5 Mechanical properties of 17-4PH and 17-7PH stainless steels.

	17-4PH	17-7PH
Elastic Modulus, ksi	28.5×10^3	29.0×10^3
Compressive Modulus, ksi	30.0×10^3	30.0×10^3
Shear Modulus, ksi	11.2×10^3	11.5×10^3
Poisson's Ratio	0.27	0.28

Table 2.6 Mechanical properties of stainless steels.

	S17400	S17700
Young's Modulus, MPa	197	200
Modulus of Elasticity in Compression, MPa	207	207
Modulus of Rigidity, MPa	77.3	79.4
Poisson's Ratio	0.27	0.28

2.9 SUMMARY

Examination of data to be put into a technical database in terms of the data types and data classes discussed above is not just an academic exercise but rather an important activity during the planning phase of a database project. The details of the planning and design stages of a technical database depend on a clear understanding of the data involved. Proper planning also identifies the access paths that need to be supported by the database and the interface required by the users.

REFERENCES

[1] *Graphics for Chemical Structure*, Ed. W. Warr, *American Chemical Society Symposium Series No. 341*, American Chemical Society, Washington DC, 1987.

[2] A. Shoshani, F. Olken, and H. Wong, Data Management Perspective of Scientific Data, in *The Role of Data in Scientific Progress, Proceed. 9th CODATA Conference*, Ed. P. G. Glaeser, North-Holland, Amsterdam, 1985.

[3] J. L. McCarthy, Information Systems Design for Materials Properties Data, in *Computerization and Networking of Materials Databases*, Eds. J. S. Glazman and J. R. Rumble, Jr., American Society for Testing and Materials, Philadelphia, PA, 1989.
J. L. McCarthy, The Automated Data Thesaurus: A New Tool for Scientific Information, in *Proceedings of the Eleventh International CODATA Conference*, Ed. P. S. Glaeser, Hemisphere Press, New York, 1990.

[4] F. J. Smith and J. G. Hughes, Some special features of a materials database system, in *Proceedings of DKSEM Conference on Data and Knowledge Systems for Manufacturing and Engineering*, Harford CT, Oct. 19-20, 1987, Computer Society of the IEEE, Washington DC 1988.

Chapter 3

Types of Technical Databases

3.1 INTRODUCTION

In this chapter technical databases are discussed from several viewpoints. We begin by looking at how databases are built during the different stages of scientific and engineering work. Each of these stages generates and handles data differently, and the resulting databases, as can be imagined, reflect the differences. We also classify technical databases according to their user community, highlighting important considerations that builders must address. Finally, we develop the concept of the **mature** database, a database that is available to a broad user group for a variety of applications.

The purpose of this discussion is to give the reader a perspective on the different types of technical databases that can be built and of the work necessary to transform one type into another. This consideration is important because often people believe that because a collection of data has been entered into a computer, delivering that data in the form of a database is a trivial process. It is not. Different technical databases have different requirements in terms of the amount and completeness of data and supporting information, the design and implementation, the type of user interface and the documentation.

3.2 THE FLOW OF TECHNICAL INFORMATION

Our first look at technical databases will be from the perspective of how they fit into the flow of technical information from its generation to its use in applications. We will develop this flow and show how the different kinds of database relate to the different stages. The reader should be able to find the reason for building a technical database in one of these stages.

The basic flow of technical information can be divided into four stages as shown in Figure 3.1. First, raw data are measured and collected either by experiments, tests,

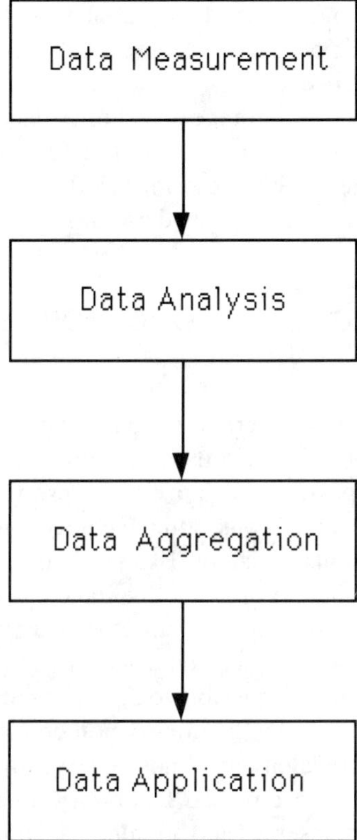

Figure 3.1 The flow of technical information.

or calculations. These results are then analyzed and reported to make them intelligible to other users. Over a period of time, reported results are aggregated together, summarized and converted to property data. Finally, data are applied to scientific and engineering problems. The data are made available to users in a wide

Table 3.1 Types of technical databases.

Source of Data	Type of Database
Measurement	Laboratory notebook
Analysis	Report
Aggregation	Handbooks
Application	Application

variety of publications and, more recently, by databases. The ideas incorporated in the flow model have been previously discussed by Kroeckel [1], Mindlin [2], Westbrook [3] and Rumble [4].

The computerized handling and storage of technical data have different functions in each stage. The databases related to these different functions (Table 3.1) also vary considerably. For each stage, we will describe briefly the technical work that goes on and discuss the different database considerations.

3.3   DATA MEASUREMENT AND LABORATORY NOTEBOOK DATABASES

Technical data begin with their generation and measurement by means of experiments, tests, calculations, and simulations. The primary function of collecting and storing these data, called **experimental** data, is the preservation of the experimental results. While this might seem obvious, some technical measurements are needed simply to support instantaneous decisions. Examples are determining the temperature of a solution to see if the next processing stage can begin or measuring the hardness of a steel to see if a shipment meets its specifications.

Many experiments or tests do yield data worth saving. It is often costly to reproduce these and sometimes impossible to do so. For many years, experimental data were collected and preserved in laboratory notebooks. Over the last 30 years, computers have become prevalent in almost every type of technical and test experiment and, in effect, have replaced laboratory notebooks. Computers can respond faster than a human observer and are more accurate; they are also capable of monitoring tests for long time periods. They certainly can collect and process large amounts of data easily.

We call the computerized collections of experimental data **laboratory notebook databases.** Though most researchers do not consider them to be databases, these data collections are usually treated in the manner of databases, being searched, analyzed, edited, updated, manipulated and displayed.

The primary features of laboratory notebook databases are the amount of data contained and their completeness. For many experiments and tests, a large amount of measured data is collected. The data consist of instrument readings, intermediate calculations, and various configuration data (see Section 2.5.1). The purpose of laboratory notebook databases is to capture in as much detail as possible all known information about a particular measurement.

One key to successful use of experimental data is the existence of supporting data often in the form of notes or comments that must be kept to make further use of the results possible. The more expensive or unique an experiment or test, the more important these data become. Such ancillary data are often well-defined, either from long practice or by standards. Consequently, from the earliest stages, researchers and research managers must be mindful that data collection is for both immediate purpose as well as long-term use. It is more difficult to make notes on a computer file than it is to write in a laboratory notebook. Either a database must have all data items or data fields for likely information that might be included or have a large free-text comments section. Free-form comments are difficult to manipulate and a well-thought-out structure is preferable.

The successful use of computers to collect test data requires two kinds of computer-related tools not usually available in laboratories: database management systems (DBMS) and data standards.

The DBMS must have the features and capabilities necessary for the collecting of test data, as discussed in Chapters 8 to 10. Data standards refer to the representation of the data themselves. How are the materials named? How many significant figures are necessary? Standards allow researchers to spend their time doing research rather than developing answers to these and similar issues which in most cases are too complicated to be solved casually.

When laboratory notebook databases are used exclusively by one person or a small research group, the normal informal procedures now prevalent in most laboratories are adequate. When the databases are to be used and distributed to a larger group of people, standards and procedures must be better defined. A database for a single person might use very abbreviated terminology, meaningful only to the individual putting in the data, and corresponding to the cryptic notes one might make in a notebook. A database used by a larger group would need conventional terminology understandable by many individuals.

Several groups have undertaken to develop standards related to the collection of technical data in databases. Figure 3.2 gives an example from surface science. Here, the format allows for the collections of very large amounts of instrument data.

We mention briefly here observational data such as the data found in satellite, weather and seismology measurements. These data collections can be very large, measured in gigabytes, and it may be years before they are analyzed. Because of the delays involved in using such data, careful documentation is needed.

In summary, laboratory notebook databases contain original measurements. If the experimental results are used solely by the generating group, only abbreviated or brief additional information needs to be stored. If results will be made available on a wider basis in the future, then complete information must be added.

3.4   DATA ANALYSIS AND REPORT DATABASES

Most experimental data must be analyzed and put into a more tractable form before being usable by others. The second stage in the flow of technical information is data analysis and reporting, i.e., the transformation of raw experimental data to a more reportable form. Analysis takes many forms, from visual checking to see if it 'looks right' to sophisticated mathematical analysis. The techniques are varied; examples are statistical analysis, correlation analysis, theoretical models, graphical analysis and fits to mathematical functions.

VAMAS Surface Chemical Analysis Standard Data Transfer Format with Skeleton Decoding Programs

W. A. Dench†, L. B. Hazell‡, M. P. Seah† and the VAMAS Community
† Division of Materials Applications, National Physical Laboratory, Teddington, Middlesex, UK TW11 0LW.
‡ BP Research Centre, Chertsey Road, Sunbury-on-Thames, Middlesex, UK TW16 7LN.

In surface analysis today many commercial instruments are operated through a computer. This computer is also used for processing the captured data, using routines from a built-in set of options for peak synthesis, peak deconvolution, background subtraction, peak area measurement, quantification in various levels of sophistication, mapping, depth profile presentation, smoothing, differentiation and a host of other functions. However, many analysts wish to process their data on another computer in their own particular way using programs written to their specification and under their full control. They need to encode the data in the data-capture computer into a form suitable for transmission then decode it into the form required in the receiving computer. A standard format for the transferred data would clearly lead to economies in the number of programs required to effect the encoding and decoding.

To meet the above requirement, as a result of a series of iterative discussions over the last three years, first within the UKESCA Users Group and then with the VAMAS National Representatives and users and manufacturers within the VAMAS Community as well as the ASTM Committee E-42.11, we have produced a Surface Chemical Analysis Standard Data Transfer Format defined in a standard metalanguage. This Format is intended to be used by those wishing to transfer data from computer to computer via parallel interfaces or via serial interfaces over direct wire, telephone line, local area network or other communications link. It is suitable for AES, EDX, FABMS, ISS, SIMS, SNMS, UPS, XPS, XRF and similar analytical methods. It covers elemental maps, depth profiles and sequences of data resulting from a variety of experiments. Thus the application of the Format is very general.

Figure 3.2 A section of the standard for reporting surface science data as developed by VAMAS.

The databases associated with the data analysis process are called **report databases** and contain the analyzed results. Today most of this data analysis is done on computers, and the use of databases to help the process and store the results is a logical consequence.

Table 3.2 Functions of technical data analysis and reporting.

Derivation of properties

Increased scientific and engineering usability

Improved understanding

Extension of data domain

Quality assurance

Uniformity

Presentation of information

There are several purposes for the data analysis and reporting process as shown in Table 3.2. In each of these cases, experimental data are examined in different ways, depending on the discipline, and the results of these actions published as journal articles, reports and other technical literature. Derived data are in a form usable in other applications. For example, a reaction rate calculated from a measured ionic current or data can be directly compared to results from other experimenters.

At the present time, few report databases exist because they generally do not contain enough data. Just as a typical published report of analyzed data contains only a few data tables and graphs, the typical report database contains only a few analyzed data sets. Generally speaking, when analyzed data are aggregated, as discussed in the next section, databases become large enough to be worth distributing.

The importance of databases in the data analysis stage cannot be minimized because of the need for preserving analysis results as well as documenting the analysis technique itself. In many cases, after the experimental measurements (i.e., ionic currents) have been converted into derived data (i.e., reaction rates), statistical analysis is performed, especially to determine the influence of certain variables. In some instances, the number of measurements of an easily changed independent variable, such as temperature, is large and the amount of data is then very large. Data reduction techniques such as curve fitting can reflect detailed variations in the experimental data while providing compact expressions for further use.

Database builders and research sponsors must become aware of the need to preserve data analysis techniques within report databases, especially so that the data are complete, the data well-characterized, the format well defined and the database well-documented. Building a database, as with a publication, costs money. This burden has to be recognized from the start and factored into schedules, milestones and budgets.

In addition, mechanisms are needed for preserving these small databases for future data aggregation. At present, data analysis results are published even though the data were initially collected on a computer, all analysis was done on a computer, and all the tables and graphs included in the publications were made on the computer. Users often need the published data in computerized form for use in simulation or modeling software and must re-enter the data and verify the accuracy. Such a process is of course subject to errors, inefficiency and incompleteness and can be improved by creating and disseminating report databases.

Some thought has been given to creating **living** databases that will collect data analysis results on an ongoing basis. To date, very few efforts have progressed beyond the discussion phase. One successful example is structure data for protein crystals. Data sets containing routine information too voluminous to publish in typical technical publications are directly contributed to a central database repository.

In summary, report databases contain the results of an analysis procedure applied to a set of test data. In some cases, the original data are included. The supporting information should include enough details on the analysis procedure that users can determine if it is acceptable or correct. Typical data resulting from analysis are properties or statistically analyzed best values.

3.5 DATA AGGREGATION AND HANDBOOK DATABASES

After data have been collected, analyzed and reported to the public, they are used for a variety of applications. Over the years, scientists and engineers have found that retrieving this type of published data is a difficult process. The data in the original literature are often not easy to use. Similar data are reported in a variety of units. Different experimental techniques have been used to measure the same properties with varying results. Data are published in a multitude of sources, often difficult to find or even to be aware of.

Consequently, data are often aggregated together with similar results in collections, usually in the form of handbook or data evaluation compilations. Aggregated data sources have become the source in many situations, thereby greatly reducing the cost and improving the efficiency of data accessibility .

A typical example of an aggregated data source is a handbook. These contain a wide range of data for large numbers of substances, and they come in a variety of formats. Handbooks can contain significant amounts of explanatory text with perhaps as much as a full page of text associated with one or two tables (Figure 3.3). Others are very compact and primarily contain tables and graphs (Figure 3.4). Usually handbook data have been evaluated to some degree and, in many instances, with individual editors responsible for small sections. These publications can also have intricate indexing schemes that reflect a need to access the large amount of data via several different paths.

Most data users are not experts in data measurement and are not particularly adept at determining the quality of data from original sources. In data compilations that are unevaluated or for which the evaluation process is not documented, determining data quality is even more difficult because experimental details have been left out. Therefore, evaluated data compilations have become very important to users.

Data evaluation represents the efforts of neutral critical evaluators who assess the quality of a given set of data regardless of the origin. Evaluation is usually done at three different levels.

(i) The original documentation reporting the data is examined for detail and assurance that all relevant parameters have been identified, controlled and taken into account. For newer technical disciplines, this can be a difficult task because influencing parameters have often not been identified. For complicated phenomena the large number of parameters may also be a problem. Through the use of standardized tests and procedures, many measurement techniques not only specify which parameters must be controlled but also set their values.

bronze), the thickness of the bearing material, and the character of the bond between the bearing material and the backing, are all factors of consequence in bearings for use in high-speed reciprocating engines, such as the main and connecting-rod bearings of automobile and aircraft engines.

Resistance to fatigue is somewhat less important in bearings that operate under more or less static load — for example, journal bearings in locomotive tenders and passenger and freight railway cars. In such bearings, antiseizure characteristics, deformability, compressive strength, and resistance to abrasion and corrosion are of greater significance. The lining metal generally employed in car and tender journal bearings (ASTM B67) conforms to the f o l l o w i n g requirements for chemical composition:

Tin, %	as specified
Antimony, min %	8.0
Tin and antimony, %	10 to 14
Arsenic, max %	0.2
Copper, max %	0.5
Sum of tin, antimony, lead and arsenic, min %	99.25
Other impurities, max %..	0.75

Pouring temperature and rate of cooling exert a profound influence on the microstructure and properties of lead-base alloys, particularly when they are used in the form of relatively heavy liners for railway journals. High pouring temperatures and slow cooling rates, such as result from the use of overhot mandrels, promote segregation and the formation of a coarse structure. A coarse structure may cause brittleness, low compressive strength, and low hardness. Low pouring temperatures (620 to 650 F) are therefore recommended. Since these alloys remain relatively fluid, almost to the point of complete solidification (about 465 F for most of these lead-base alloys), they are easy to manipulate and can be handled with no great loss of metal from drossing.

The use of the lead-base babbitts containing calcium and the alkaline earth metals is confined almost entirely to the railway field, although these babbitts are employed to some extent also in certain diesel engine bearings.

One of the most widely used of these alloys contains 1.0 to 1.5% Sn, 0.50 to 0.75% Ca, and small amounts of various other elements. The strength of this alloy approximates that of the alloy that contains 90% Sn, 8% Sb and 2% Cu, referred to in Table 3. The Brinell hardness is about 20, and the solidus is 610 F. The liquidus is probably near 640 F. The pouring temperature, which varies from 930 to 970 F, is relatively high and accounts for the formation of a larger volume of dross than is encountered when melting the lead-base Pb-Sb-Sn alloys. Care must be taken to avoid contamination of the alloy by antimonial lead-base babbitts, and vice versa. Deformability and resistance to wear are of the same order as those of the other lead-base babbitts. Most alloys of this type are subject to corrosion by acidic oils.

Copper-Lead Alloys

Copper-lead alloys are used extensively in automotive and aircraft applications and in general engineering.

In the aircraft and automotive fields, these alloys are used almost exclusively in steel-backed bearings, a majority of which (for the automotive field) are blanked from steel strip continuously coated with alloy in automatically controlled machines. This same general method of manufacture is used in the production of most steel-backed babbitt bearings. Typical alloys are SAE 48, 49 and 480; their compositions are listed in Table 7.

SAE 48 was designed originally for aircraft engine bearings. It is now used in automobiles. SAE 480 is used exclusively in general automotive practice. The tolerance for lead in SAE 480 has been purposely made broad enough to cover the several compositions now in commercial use. Closer tolerances may be specified by the purchaser of the alloy.

Copper-lead a l l o y s containing about 30% Pb and as much as 5% Ag have been used in the automotive industry. Another type of bearing has an intermediate layer of 25% Pb copper-lead cast on a steel back, and a thin plated overlay of lead-tin-copper alloy. These alloys have exceptionally high resistance to fatigue and are widely used in the automotive, diesel, and aircraft fields.

The higher the lead content of the copper-lead alloys, the lower the fatigue strength and the higher the antifriction characteristics. Silver and tin may be substituted for part of the copper in order to increase fatigue strength. These alloys are easily corroded by certain oils. In fact, corrosion can be a major problem with bare copper-lead alloys under certain conditions. Overlay-plated copper-leads, containing 8 to 12% Sn in the lead overlay, are sufficiently corrosion resistant to survive corrosive oil conditions. A nickel or brass dam, approximately 50 micro-in. thick, is necessary to prevent the diffusion of the tin to the copper phase of the intermediate layer. In the absence of this dam, the babbitt overlay soon loses the protection of the tin in the lead.

Copper-lead bearings are also made by applying copper-lead powder to steel strip that is passed continuously through furnaces where the mixtures are simultaneously sintered and bonded to the strip. In these bearings, the structure is homogeneous and equiaxed, whereas in cast copper-lead strip bearings, the copper dendrites are essentially perpendicular to the surface of the strip.

In general engineering, alloys of copper and lead have found wide application as bearings and bushings. The hardness and the compressive and tensile strengths of these alloys decrease as the lead content is increased. Table 8 presents approximate ranges for the mechanical properties of sand cast bearings of these materials.

The properties of these alloys are profoundly affected by the method of manufacture and by the size of the section cast. They have good antiseizure characteristics and can be used with some success where lubrication is uncertain or intermittent. Although copper-lead bearings are used without special support in some installations, it is general practice in the automobile field to apply these alloys as a thin layer on a steel shell.

Because lead may segregate during slow cooling, as in sand casting, tin, nickel, manganese and other elements are added to reduce segregation. These elements increase the strength of the basic alloy and provide materials that are used extensively in engineering.

Tin Bronzes

The bearing bronzes are substantially alloys of copper, with from 5 to 20% Sn, and a small percentage of residual phosphorus — phosphorus remaining after deoxidation of the alloys with 10 or 15% phosphor copper. These alloys are sometimes referred to as "phosphor bronzes".

When 2 to 6% Zn is used instead of phosphorus, the alloy is quite generally referred to as "gun metal" (also, as Government bronze, composition G, admiralty bronze, or zinc bronze).

Table 7. Compositions of Copper-Lead Bearing Alloys Cast or Sintered on Steel

SAE alloy	Cu	Pb	Ag, max	Zn, max	P, max	Fe, max	Sn, max	Others, max
49	73 to 79	21 to 27	....	....	....	0.35	0.50	0.45
48	67 to 74	25 to 32	1.5	0.1	0.025	0.35	0.25	0.15
480	60 to 70	30 to 40	1.5	....	....	0.35	0.50	0.3

Table 8. Mechanical Properties of Sand Cast Copper-Lead Bearings Containing 30 and 40% Pb

Property	30% Pb	40% Pb
Tensile strength, psi	8000 to 9000	7500 to 8500
Elongation, % in 2 in.	6 to 8	6 to 8
Compressive strength(a), psi	3000	2500
Brinell hardness number	22.5 to 32.5	20 to 30

(a) 0.001 in. compression on sample 1 in. high and 1 in. in diameter

---

Figure 3.3 A page from a typical handbook with much text explaining one table.

PHYSICAL CONSTANTS OF ORGANIC COMPOUNDS (Continued)

No.	Name, Synonyms, and Formula	Mol. wt.	Color, crystalline form, specific rotation and λ_max (log ε)	b.p. °C	m.p. °C	Density	n_D	Solubility	Ref.
6637	Ethane, 1,2-diido or Ethylene diiodo CH_2ICH_2I	281.86	ye mcl pr or rh (eth) d in lt	200, 74[10]	83	3.325[20/4]	1.871[20]	al, eth, ace, chl	B1[4], 169
6638	Ethane, 1,2-di-N-morpholyl $C_{10}H_{20}N_2O_2$	200.28	wh-yesh (eth or lig)	160-3[25]	75			w, al, ace, bz	B27, 7
6639	Ethane, 1,1-dinitro $CH_3CH(NO_2)_2$	120.07	ye mcl (bz or MeOH)	185-6, 72[12]		1.3503[14/24]		al, eth	B1[4], 174
6640	Ethane, 1,2-dinitro-1,1,2,2-tetrafluoro $O_2NCF_2CF_2(NO_2)$	192.03		58-9	−41.5	1.6024[25/4]	1.3265[25]	ace	B1[4], 175
6641	Ethane, 1,1-diphenyl or *a*-Methylditan $CH_3CH(C_6H_5)_2$	182.27		286, 148[15]	−21.5	0.9997[20/4]	1.5756[20]	al, eth, bz	B5[4], 1880
6642	Ethane, 1,2-diphenyl or Bibenzyl $C_6H_5CH_2CH_2C_6H_5$	182.27	nd (al)	285, 95-6[1]	52.2	0.9583[60/4]	1.5478[60]	al, eth	B5[4], 1868
6643	Ethane, 1,2-di-(4-tolyl) $(4\text{-}CH_3C_6H_4)CH_2CH_2(C_6H_4CH_3\text{-}4)$	210.32	lf(MeOH or dil al) pl (lig)	296-8, 178[18]	82-3			bz, peth	B5[4], 1943
6644	Ethane, 2,2-di-4-tolyl-1,1,1-trichloro $(4\text{-}CH_3C_6H_4)_2CHCCl_3$	313.56	mcl pr (al eth-al)		92			al, eth, ace	B5[4], 1949
6645	Ethane, 1-ethoxy-2-methylamino or Ethyl-β-methyl amino ethyl ester $C_2H_5OCH_2CH_2NHCH_3$	103.17		114-5[744]		0.8363[20/4]	1.4147[20]	w, al, eth, ace, bz	B4[3], 647
6646	Ethane, fluoro or Ethyl fluoride CH_3CH_2F	48.06	gas	−37.7	−143.2	0.7182[20/4] (liq)	1.2656[20]	al, eth	B1[4], 120
6647	Ethane, 1-fluoro-1,2,2-trichloro $Cl_2CHCHFCl$	151.40		101-3		1.54968[17]	1.4390[20]D		B1[4], 141
6648	Ethane, fluoro penta chloro FCl_2CCCl_3	220.29		134-6	101.3			al, eth	B1[4], 148
6649	Ethane, hexabromo or Perbromo ethane C_2Br_6	503.48	rh pr (bz)	d200-10	d	2.823[20/4]	1.863		B1[3], 193
6650	Ethane, hexachloro or Perchloro ethane C_2Cl_6	236.74	rh (al-eth)	186[777]	186-7 (sealed tube)	2.091[20/4]		al, eth, bz	B1[4], 148
6651	Ethane, hexafluoro or Perfluoro ethane C_2F_6	138.01	gas	−79	−94	1.590[-78]			B1[4], 123
6652	Ethane, hexaphenyl $(C_6H_5)_3CC(C_6H_5)_3$	486.67	cr (ace)	d	145-7d			eth, ace, chl, MeOH	B5[3], 2746
6653	Ethane, iodo or Ethyl iodide C_2H_5I	155.97		72.3	−108	1.9358[20/4]	1.5133[20]	al, eth	B1[4], 163
6654	Ethane, isocyano or Ethyl carbylamine CH_3CH_2NC	55.08		79[775]	<−66	0.7402[20/4]	1.3622[20]	al, eth, ace	B4[4], 342
6655	Ethane-1-(4-methoxyphenyl)-1-phenyl or 1-*p*-Anisyl-1-phenyl ethane $CH_3CH(C_6H_5)(C_6H_4OCH_3\text{-}4)$	212.30		180-2[19]		1.0473[20/4]	1.5725[20]	eth, ace, bz	B6[3], 639
6656	Ethane, nitro $C_2H_5NO_2$	75.07		115	−50	1.0448[25/4]	1.3917[20]	al, eth, ace	B1[4], 170
6657	Ethane, nitro-pentafluoro $CF_3CF_2NO_2$	165.02		0				eth	B1[4], 172
6658	Ethane, 1,nitro-2,2,2-trifluoro $F_3CCH_2NO_2$	129.04		96		1.3914[20/4]	1.3394[20]	eth	B1[4], 172
6659	Ethane, nitroso-pentafluoro CF_3CF_2NO	149.02		−42					B1[4], 169
6660	Ethane, pentabromo Br_2CHCBr_3	424.58	mcl pr (dil al)	210[100]	56-7	3.312[20/4]		al, eth	B1[3], 193
6661	Ethane, pentachloro $CHCl_2CCl_3$	202.30		162	−29	1.6796[20/4]	1.5025[20]	al, eth	B1[4], 147
6662	Ethane, pentaiodo CHI_2CI_3	659.55	mcl pr (aa)		182-4			al, eth, bz, aa	B1[3], 31
6663	Ethane, perfluoro CF_3CF_3	138.01		−79	−100.6	1.590[-98]			B1[4], 123
6664	Ethane, 1,1,1,2-tetrabromo CH_2BrCBr_3	345.67		112[19],	0.0	2.8748[20/4]	1.6277[20]	al, eth, ace, bz, chl	B1[4], 162
6665	Ethane, 1,1,2,2,-tetrabromo $CHBr_2CHBr_2$	345.67	yesh	243.5, 114.8[10]	0	2.9656[20/4]	1.6353[20]	al, eth, ace, bz, aa	B1[4], 162
6666	Ethane, 1,1,1,2-tetrachloro CH_2ClCCl_3	167.85	yesh red	130.5, 22.1[10]	-70.2	1.5406[20/4]	1.4821[20]	al, eth, ace, bz, chl	B1[4], 143
6667	Ethane, 1,1,2,2-tetrachloro $CHCl_2CHCl_2$	167.85		146.2, 33.9[10]	−36	1.5953[20/4]	1.4940[20]	al, eth, ace, bz	B1[4], 144

Figure 3.4 A page from a handbook with dense tables and graphs.

(ii) The data behaviour is compared to known physical laws that govern the phenomena involved. Many technical measurements are analyzed and transformed into property data which should obey these laws. For example, thermochemistry data must follow the laws of thermodynamics. Often the laws are known only empirically and data not obeying them must be examined carefully to determine if they are incorrect or if the empirical relationship is failing, for example, the tribology data mentioned in Chapter 2.

(iii) Data are compared to other measurements of the same quantity. Many properties can be measured using several independeht techniques. Intercomparison of results from different methods can identify suspect data. An example of this is the network of chemical thermodynamic data which allows for testing the consistency of a new data value with all other previous measurements.

Data evaluators employ combinations of these three approaches depending on the technical area. Evaluation needs to be performed by experts since their opinions are most respected. Data are also evaluated by research groups, technical committees, data centres and boards of editors and authors. Such collective actions are usual for engineering data.

Data compilations covering all data on a given subject can also be analyzed over and above the analysis of a single test. This technique has been very successful in identifying suspect measurements. For example, large amounts of crystallographic data have been studied, and it has been found that certain crystal classes of the 230 possible classes have only one or very few naturally occurring examples (Figure 3.5). The scarcity of occurrences indicates that these compounds might exhibit interesting chemical features or might simply result from bad measurements.

An important type of technical database corresponds to those data compilations which are called **handbook** databases. These generally contain a wide range of data for a number of substances and represent a compilation and selection of available analyzed data. Some correspond to printed data collections such as the databases issued by the Standard Reference Data Program of the National Institute of Standards and Technology (formerly the National Bureau of Standards) that correspond to data compilations printed in the *Journal of Physical and Chemical Reference Data.*

Handbooks and other data compilations are usually the data source of first choice. Unfortunately, few handbook databases now exist, and it is a void that is keenly felt. However, some technical areas exist that are exceptions. One is chemistry where several online networks allow access to different combinations of chemical databases that in effect act as handbooks. This subject is discussed further in the next chapter.

Not all handbook databases contain evaluated data. For example, companies are now producing product databases that primarily contain information on their products. The data might be wide in coverage though they are simply a collection of measurements made by the company. These product databases are similar to published data sheets and often provide important summary information to product selectors, for example, for component selection in computer-aided design. However, the data included in these databases reflect the needs of the producing company and not of the consuming scientific public. Data can be issued primarily to build sales or make products look attractive. Database users must be aware of their purpose.

Space-group frequencies for 29 059 organic crystalline compounds

Space-group symbol	Space-group No.	Frequency	Space-group symbol	Space-group No.	Frequency	Space-group symbol	Space-group No.	Frequency	Space-group symbol	Space-group No.	Frequency
$P1$	1	305	$Pmmn$	59	23	$P\bar{4}b2$	117	2	$P6/m$	175	1
$P\bar{1}$	2	3986	$Pbcn$	60	341	$P\bar{4}n2$	118	4	$P6_3/m$	176	75
$P2$	3	11	$Pbca$	61	1261	$I\bar{4}m2$	119	1	$P622$	177	2
$P2_1$	4	1957	$Pnma$	62	548	$I\bar{4}c2$	120	2	$P6_122$	178	6
$C2$	5	273	$Cmcm$	63	61	$I\bar{4}2m$	121	12	$P6_522$	179	1
Pm	6	1	$Cmca$	64	96	$I\bar{4}2d$	122	22	$P6_222$	180	4
Pc	7	102	$Cmmm$	65	4	$P4/mmm$	123	0	$P6_422$	181	1
Cm	8	22	$Cccm$	66	12	$P4/mcc$	124	8	$P6_322$	182	6
Cc	9	277	$Cmma$	67	2	$P4/nbm$	125	4	$P6mm$	183	0
$P2/m$	10	5	$Ccca$	68	14	$P4/nnc$	126	1	$P6cc$	184	0
$P2_1/m$	11	239	$Fmmm$	69	3	$P4/mbm$	127	2	$P6_3cm$	185	1
$C2/m$	12	189	$Fddd$	70	30	$P4/mnc$	128	14	$P6_3mc$	186	15
$P2/c$	13	141	$Immm$	71	4	$P4/nmm$	129	19	$P\bar{6}m2$	187	0
$P2_1/c$	14	10450	$Ibam$	72	27	$P4/ncc$	130	16	$P\bar{6}c2$	188	0
$C2/c$	15	1930	$Ibca$	73	8	$P4_2/mmc$	131	3	$P\bar{6}2m$	189	0
$P222$	16	7	$Imma$	74	5	$P4_2/mcm$	132	1	$P\bar{6}2c$	190	9
$P222_1$	17	9	$P4$	75	1	$P4_2/nbc$	133	2	$P6/mmm$	191	1
$P2_12_12$	18	187	$P4_1$	76	47	$P4_2/nnm$	134	2	$P6/mcc$	192	7
$P2_12_12_1$	19	3359	$P4_2$	77	3	$P4_2/mbc$	135	1	$P6_3/mcm$	193	0
$C222_1$	20	86	$P4_3$	78	7	$P4_2/mnm$	136	17	$P6_3/mmc$	194	9
$C222$	21	5	$I4$	79	12	$P4_2/nmc$	137	8	$P23$	195	0
$F222$	22	0	$I4_1$	80	9	$P4_2/ncm$	138	3	$F23$	196	0
$I222$	23	7	$P\bar{4}$	81	7	$I4/mmm$	139	17	$I23$	197	3
$I2_12_12_1$	24	5	$I\bar{4}$	82	59	$I4/mcm$	140	4	$P2_13$	198	15
$Pmm2$	25	2	$P4/m$	83	6	$I4_1/amd$	141	11	$I2_13$	199	0
$Pmc2_1$	26	12	$P4_2/m$	84	3	$I4_1/acd$	142	19	$Pm3$	200	2
$Pcc2$	27	0	$P4/n$	85	37	$P3$	143	10	$Pn3$	201	0
$Pma2$	28	1	$P4_2/n$	86	48	$P3_1$	144	21	$Fm3$	202	0
$Pca2_1$	29	242	$I4/m$	87	28	$P3_2$	145	10	$Fd3$	203	1
$Pnc2$	30	3	$I4_1/a$	88	98	$R3$	146	40	$Im3$	204	3
$Pmn2_1$	31	37	$P422$	89	1	$P\bar{3}$	147	26	$Pa3$	205	36
$Pba2$	32	9	$P42_12$	90	4	$R\bar{3}$	148	122	$Ia3$	206	5
$Pna2_1$	33	513	$P4_122$	91	3	$P312$	149	0	$P432$	207	0
$Pnn2$	34	14	$P4_12_12$	92	101	$P321$	150	5	$P4_232$	208	0
$Cmm2$	35	2	$P4_222$	93	2	$P3_112$	151	0	$F432$	209	1
$Cmc2_1$	36	56	$P4_22_12$	94	7	$P3_121$	152	27	$F4_132$	210	3
$Ccc2$	37	6	$P4_322$	95	1	$P3_212$	153	0	$I432$	211	0
$Amm2$	38	0	$P4_32_12$	96	44	$P3_221$	154	8	$P4_332$	212	1
$Abm2$	39	5	$I422$	97	2	$R32$	155	23	$P4_132$	213	1
$Ama2$	40	14	$I4_122$	98	1	$P3m1$	156	1	$I4_132$	214	0
$Aba2$	41	47	$P4mm$	99	0	$P31m$	157	4	$P\bar{4}3m$	215	7
$Fmm2$	42	8	$P4bm$	100	0	$P3c1$	158	3	$F\bar{4}3m$	216	1
$Fdd2$	43	115	$P4_2cm$	101	0	$P31c$	159	5	$I\bar{4}3m$	217	18
$Imm2$	44	3	$P4_2nm$	102	4	$R3m$	160	21	$P\bar{4}3n$	218	6
$Iba2$	45	31	$P4cc$	103	0	$R3c$	161	39	$F\bar{4}3c$	219	2
$Ima2$	46	5	$P4nc$	104	3	$P\bar{3}1m$	162	0	$I\bar{4}3d$	220	4
$Pmmm$	47	4	$P4_2mc$	105	0	$P\bar{3}1c$	163	13	$Pm3m$	221	3
$Pnnn$	48	3	$P4_2bc$	106	1	$P\bar{3}m1$	164	15	$Pn3n$	222	0
$Pccm$	49	1	$I4mm$	107	2	$P\bar{3}c1$	165	17	$Pm3n$	223	5
$Pban$	50	2	$I4cm$	108	1	$R\bar{3}m$	166	20	$Pn3m$	224	1
$Pmma$	51	9	$I4_1md$	109	6	$R\bar{3}c$	167	36	$Fm3m$	225	22
$Pnna$	52	23	$I4_1cd$	110	9	$P6$	168	0	$Fm3c$	226	0
$Pmna$	53	15	$P\bar{4}2m$	111	1	$P6_1$	169	14	$Fd3m$	227	1
$Pcca$	54	13	$P\bar{4}2c$	112	0	$P6_5$	170	16	$Fd3c$	228	4
$Pbam$	55	12	$P\bar{4}2_1m$	113	17	$P6_2$	171	5	$Im3m$	229	8
$Pccn$	56	101	$P\bar{4}2_1c$	114	68	$P6_4$	172	0	$Ia3d$	230	0
$Pbcm$	57	64	$P\bar{4}m2$	115	1	$P6_3$	173	33			
$Pnnm$	58	30	$P\bar{4}c2$	116	0	$P\bar{6}$	174	1			

Figure 3.5 The distribution of crystal structures among the 230 possible space groups. As can be noted, some classes have many occurrences while others have few or none. Data for those classes with only a few compounds can be re-examined for error or for possible unique properties.

3.6   DATA APPLICATIONS AND APPLICATIONS DATABASES

The work of a scientist or engineer involves problem-solving as well as research. Rapid access to pertinent data is a major factor in determining solutions in a timely fashion. A data collection targeted to one specific application area containing relevant data from a wide range of sources is called an **applications database**. In the future, when technical information is completely computerized, applications databases will be of primary importance.

Over the years scientists and engineers have found it convenient to create specific data collections for individual applications. They do this (i) for convenience because the work they do often requires data from a wide variety of different sources and (ii) for quality because the data have already been changed to meet their needs. Data in these specialized collections are often much changed from the original measurements and represent the highest refinement that technical data undergo.

Applications databases are now being built for specific applications, directly comparable to the specialized publications mentioned above. The decisions taken in building such databases are entirely dependent on the application. The user interface and the search and display strategies are optimized to the application.

In other circumstances, applications databases will be put together from different data sources, both published and in computer form along with new data directly related to the particular problem. Consider the following examples.

(i) A biologist might be called upon to determine the cause of a fish kill near a sewage plant outlet. Monitoring data from the outlet might be taken from a database maintained by the plant and then data from fish autopsies and water quality measurements added.

(ii) A materials specialist might be called upon to determine the structural integrity of bolts and other fasteners within a nuclear power plant. Data must be taken from the design, plant performance and materials databases, then combined perhaps with new property test results.

These working databases often take on a value and life of their own in the sense that they remain after a project is completed. The more intense the work and the longer it takes, the more likely it is that such a database will become important. When its value is recognized, steps must be taken to preserve it and expand its use. Often this would require additional resources or time that cannot be easily justified. Critical decisions must then be made regarding the future of the database. An analogous situation exists with respect to paper data collections. Often these are put in files never to be used again because of the high cost of cleaning up the data or adding full documentation. The same applies to databases if they are archived to a tape library.

One of the most expensive and difficult aspects of building a database is retrofitting, i.e., adding or changing information. Databases intended for short-term projects should be reviewed at the earliest possible stage for possible preservation or long-term use. Preservation can be achieved, but not without the type of planning and resources required to change a talk into a journal article. It does not just happen. The decision to change the nature of such a database must be made consciously and with careful planning.

Many applications databases, namely those that are becoming successful, are products of well-planned and deliberate efforts to appeal to a given market. The time and effort that have gone into them are considerable, but their developers have made conscious decisions and know their goals. Other application databases that result from wishful thinking and lack adequate support are failures. If resources are not readily available, database builders need seriously to consider stopping the project before wasting time and money.

3.7 DATABASES AND THE FLOW OF TECHNICAL INFORMATION — A SUMMARY

We have discussed how technical databases can be built in the normal course of the creation and use of technical data. At each stage in the flow of technical information, databases can logically and easily be built but for different reasons and with different characteristics and for different types of use (Figure 3.6). Databases

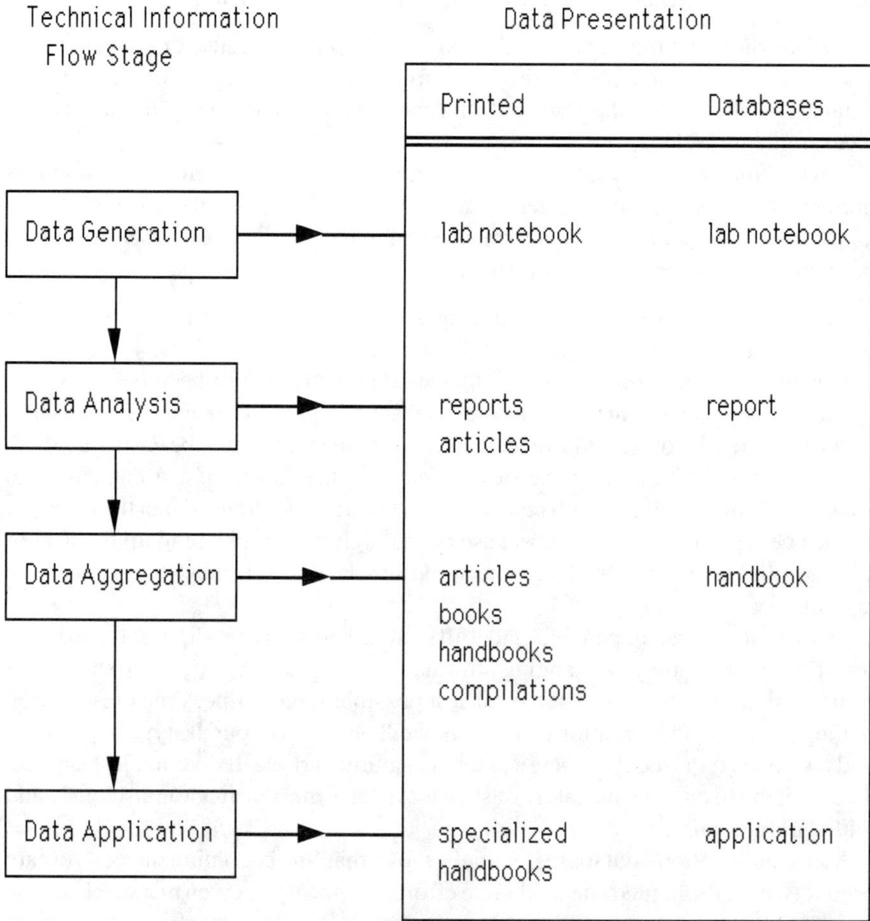

Figure 3.6 Publications and databases arising in different stages of the flow of technical information.

arising in one stage may be inappropriate for use in another stage. People needing data for one purpose may find that databases created at a different stage have too much or too little information or are just too cumbersome.

A problem that has not been addressed by technical database builders is how data might flow through the system. In the future this will be a major consideration because, as we have indicated, computers have taken over all aspects of technical work. Report databases do not now get their data from laboratory notebook databases. Handbook database builders certainly do not extract data from report databases.

Readers should now be in a position to classify their database efforts with respect to the flow of technical information and to assess the possibility that the database will cross from one stage to another. Careful planning is needed to make sure that the database will support its use at the needed level (see Chapter 10). Search paths, data items included and output displays all change from stage to stage and need to be reviewed carefully.

3.8 CLASSIFICATION OF TECHNICAL DATABASES BY TYPES OF USERS

A second classification scheme for technical databases can be made with respect to their user community as given in Table 3.3.

A **personal database** is intended only for its creator. Its use may be intensive or sporadic. Because it is aimed at only the builder, many short cuts and abbreviations may be used, depending on the memory or habits of the builder.

A **group database** is one used by a group working together on the same problem or using the same experimental equipment or computer software. The users are normally geographically associated, even if by telecommunications. Again, the contents are characterized by their brevity and the informality of conventions and documentation. Depending on the size and closeness of the group involved, these will still be more formal than for a personal database.

Table 3.3 Classification of scientific and technical databases by types of users.

Personal (one person)
Group
Institutional
Collegial
Public
Archival

When a database becomes an **institutional database**, a different level of support is involved, and more formal conventions are needed. Included are databases used by several groups, by a company, or even by a large corporation. At this level, good formal documentation is needed and careful planning and design are important to accommodate multiple needs. However, some conventions are still likely, reflecting common institutional practice. For example, materials or chemicals might be referred to by trade names.

A **collegial database** is one used across institutions, by both small and large numbers of people working in a related technical area, but usually on a fairly formal basis. The data contents may use more general terminology, thereby avoiding trade-name problems or proprietary concerns. Collegial databases are often associated with large-scale modeling efforts or observational data-collection efforts. Formats for data contributors may be well-defined. Documentation quality can vary, but the larger the community of colleagues, the more extensive it will be.

Public databases are those made available to the public or to a significant portion thereof. Since these databases are often used differently than anticipated or intended, documentation needs to be complete, and if wide usage is intended, the contents should not use limiting or iconoclastic terminology.

Archival databases place primary emphasis on managing and saving data for future use. Sometimes a need for these arises from the sheer volume of data. For example, gigabytes of data might be produced from satellite observations. Data may also be archived because the immediate demand for them has passed but a future demand is foreseen. For example, engineering data for advanced energy conversion is being saved for use during the next energy crisis.

Finally, data are often archived because their immediate use has been completed, but further analysis is anticipated in the future, for example, seismic or weather observations. These large volumes of data require database management techniques mainly beyond the scope of this book [5]. The advent of optical disks and other high-density storage devices will help (see Chapter 7). Major considerations for data archiving lie in the area of documentation, obsolescence of storage media and loss of corporate memory.

The level of usage is as important as the stage of technical information flow in characterizing a technical database. Moving a database from one level of usage to the next is often accomplished only with difficulty if done without planning. The mere existence of a group database does not imply that it can be used by the institution as a whole or be distributed to colleagues. References, documentation and metadata may have to be retrofitted, and this can be one of the most expensive and time-consuming acts related to database building. This is why planning and designing all databases is so important: to identify whether more widespread use is probable and, if so, to take this into account from the beginning. After consideration, it may be decided that indeed such wider use will not occur or is not worth the expected effort and/or expense. Planning may take time, but usually only a few days—a meager cost compared to the time spent retrofitting a database.

3.9   MATURE DATABASES

Mature databases have been defined by Rumble [4] as databases that have the attributes listed in Table 3.4.

Of the different levels of databases discussed above, institutional, collegial and public databases all can attain maturity, but when these databases fail to meet one or more of the standards set out in Table 3.4, they can be expected to get little use. This has been shown to be the case many times. Database builders need to recognize that their planning and design change distinctly when a mature database is desired.

First, a mature database must be complete to an advertised criterion, must contain substantial information and must perform as advertised. If it is a database on birds, it should contain data on all or most types of birds. If it is a database only on a specific type of birds, i.e., songbirds, it should have data on most songbirds.

The documentation for a mature database should be self-explanatory, and further interaction between users and builder should not be necessary. Finally, a mature database should be intended for a user community besides that of the builder.

The concept of a mature database has been developed to help distinguish between databases actually available and those which are being planned or which are not completed. Directories of technical databases should only contain mature databases, and users should insist that only databases passing all these tests of maturity be accepted.

Table 3.4 Attributes of a mature technical database.

Intended for use by other than builder
Complete, according to some criteria
Documented
Available to the public
In usable form

3.10   BARRIERS TO BUILDING TECHNICAL DATABASES

The above discussion provides the basis for understanding the types of technical databases. Different databases are built for different reasons and will have different data, metadata, distribution and documentation. Building technical databases has not proceeded as quickly as many people have predicted, in some cases because the technical building process was difficult and poorly understood. Equally important in slowing progress are several frequently overlooked social and economic factors (Table 3.5) [6].

Table 3.5 Socio-economic barriers in building technical databases.

High cost of data entry
Difficulties in database building
Lack of obvious economic benefits
Small number of well-articulated demands
Lack of encouragement from major on-line vendors

The last three barriers are changing rapidly. Industry has begun a major effort to computerize as many activities as possible. Accessing data via computer is now being recognized as a key part of that process. In addition, there is increased awareness of the importance of having the best possible data to use in making technical decisions. The distribution of technical data via the existing paper system is impressive but is not as good as it could be. For example, a large part of a design engineer's time is taken up searching for information on components and materials, using catalogues, brochures and manuals. This information should be available in seconds on the same screen used for computer-aided design.

However, the first two barriers are still problems. Presently, with only a few exceptions, traditional sources of support for scientific and technical work have not recognized the need to make funds available for putting data into computerized form. Building technical databases costs money, far more than people envisioned a few years ago. Data collection is costly, even if it means collecting already-published data. Data entry is very expensive, especially because of the need to verify the input. The database design and building process requires more effort than imagined, even with the enhanced tools now available. Finally, the design and implementation of a finished product, a mature database, is expensive.

This has not deterred a number of information vendors, both traditional and newcomers, from becoming involved in database developments, and today's increased activity reflects that interest. Very few technical databases have received wide circulation, and as of 1990, the best sellers generally number their successes in the high hundreds. Databases that have been incorporated into analytical instruments have received very high distribution. Some mass spectra databases have well over 10 000 copies in use.

Several harsh economic facts stare database builders in the face and must be considered. The price of databases is high, very high today in comparison to corresponding printed publications. With the explosion of personal computer databases, distribution problems similar to those for books and journals are rising. Heavy advertising will be needed. There are so many databases and books that no-one can be aware of all of them or can buy all those which are pertinent. It is difficult for companies to justify many purchases of similar databases or books. Finally, databases lack compatibility, thereby requiring more training and more time to use than books.

Today's technical databases offer data manipulation capabilities that were un-imagined just 10 years ago; these alone can be worth many times the cost of the database. Visual displays of chemical structure information are beginning to take on the beauty and clarity of many video games, with much higher pay-offs. One successful new chemical or pharmaceutical will pay for a database hundreds or thousands of times over.

Online systems will have an important place in the economics of database dissemination because they will provide essential support to corporate-wide information systems and will provide convenient access to a multitude of data sources. But it is far too premature to make many predictions: just five years ago there were probably only 25 databases of factual technical data charging money for their use. Today, though in the hundreds, these types of databases are still fewer than the tens of thousands of printed data sources. Overcoming the socio-economic barriers is an important factor in making progress.

3.11 SUMMARY

Technical databases have been discussed from several points of view relative to the flow of technical information with the intent of classifying them in terms of where they are built and who uses them. In determining the course of a technical database project, both considerations are important and deeply affect the planning and the actual building processes. The builders and users who understand how their technical databases fit into the overall flow of technical information will be well-prepared for success.

REFERENCES

[1] E. Bullock, H. Kroeckel, M. van de Voorde, Data Systems for Engineering Materials, the Materials Engineer's point of view, in *Materials Data Systems for Engineering*. A CODATA workshop, J. H. Westbrook et al, Eds., Fachinformationszentrum Energie, Physik, Mathematik GmbH, Karlsruhe, FRG, 1986.

[2] H. Mindlin and S. H. Smith, *Database System Considerations in Engineering Design, in Managing Engineering Data: The Competitive Edge*, Ed. R. E. Fulton, American Society of Mechanical Engineers, New York, 1987.

[3] J. H. Westbrook, Some considerations in the Design of Properties Files for a Computerized Materials Information System, in *The Role of Data in Scientific Progress*, Ed. P. S. Glaeser, North-Holland, Amsterdam, 1985.

[4] J. Rumble, J. Sauerwein and S. Pennell, *Scientific and Technical Factual Databases for Energy Research and Development*, US Dept of Energy, Office of Scientific and Technical Information, Oak Ridge, TN, No DOE/TC.40017-1, 1986.

[5] See for example any of the Proceedings for the Conference on Very Large Databases.

[6] J. R. Rumble, Jr., Socioeconomic Barriers in Computerizing Materials Databases, in *Computerization and Networking of Materials Data Bases*, J. S. Glazman and J. R. Rumble, Jr., Eds., American Soc. for Testing and Materials, Philadelphia, 1989.

Chapter 4

The Use of Technical Databases

4.1 INTRODUCTION

A major thesis of this book is that databases should be user-driven. The work of scientists and engineers is to solve problems; consequently, we must understand how databases are used to aid problem-solving. In this chapter, we will discuss the primary uses made of technical databases and show how different uses change the demands on technical database design and building. Typical uses of technical databases are shown in Table 4.1. Each involves different types of requests, different kinds of searches and different outputs.

Table 4.1 Typical uses of technical databases.

1. Searching for specific data items or data values

2. Downloading and transferring data

3. Data display

4. Analyzing data

5. Generating publications and reports

6. Looking up tables and graphs

4.2 SEARCHING FOR DATA VALUES

A technical database is most commonly used for data searching, either through standardized queries or via user-defined queries. Because most technical data are the properties of specific or general materials, the two most common access paths or search strategies are:

(i) What substances have a given set of properties?
(ii) What properties does a given substance have?

There are many variations on these questions, and some will be discussed later. Most technical databases now being disseminated emphasize standard queries; an example is given in Figure 4.1. Their advantage is that the amount of database indexing is reduced, and the size and access speed for the database are also reduced. The user-interface design becomes easier, and the need to resort to complicated and structured query languages or selection procedures is eliminated.

```
              NIST XPS Database Main Menu


   Options for Database Use

      Identify:


          1. one unknown spectral feature
          2. a set of unknown spectral features


   Search and Display for an element:


          3. binding energy data
          4. Auger kinetic energy data
          5. chemical shifts


      Browse:


          6. individual data fields


   Exit 7.


      Please enter choice >
```

Figure 4.1 The opening menu for the NIST X-Ray Photoelectron Spectroscopy Database. Users may specify a predetermined search strategy (selections 1 to 5) or set up their own (selection 6).

Defining all possible search requests may not be easily accomplished and, since the number of data fields for some technical databases is high, sometimes in the hundreds, user choice in formulating queries is a real benefit. A recently designed x-ray photoelectron spectroscopy database takes this into account by providing a small number of standard access paths plus a menu for user-specified queries as shown on the screen in Figure 4.1. If the user chooses the option 6 in Figure 4.1, the screen in Figure 4.2 comes up, allowing range-searching on arbitrarily chosen data fields.

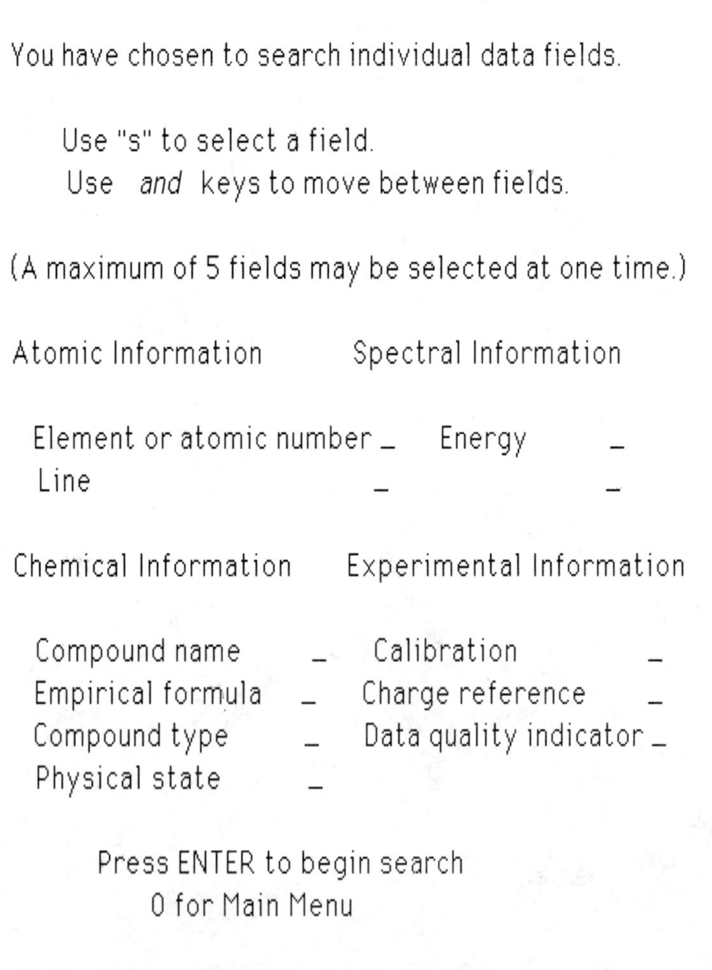

Figure 4.2 User specification menu for NIST X-Ray Photoelectron Spectroscopy Database. The user can choose any combination of data fields.

An alternative approach using windows is shown in Figure 4.3. The selections in the left-hand window represent major categories of corrosion data fields. The left-centre window gives subcategories for the choice material.

4.2.1 Searching for chemicals, engineering materials, and other substances

For property databases, the description of the substance, whether a chemical, an engineering material, a bird or a plant, for which properties are available is the single most important feature. Unfortunately, different technical disciplines have developed very elaborate and highly overlapping systems of nomenclature and designation. For example, chemists can visualize chemicals and communicate their understanding to their colleagues through many different systems. The possibilities (Figure 4.4) include chemical formulae, chemical names, two- and three-dimensional geometrical models and atomic connection tables.

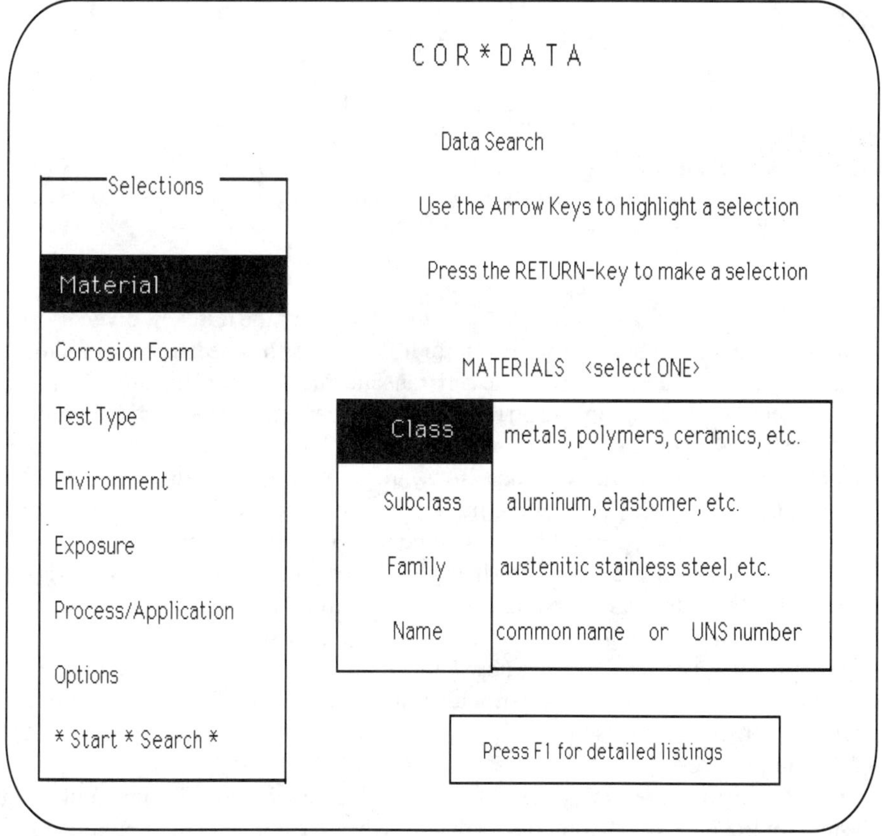

Figure 4.3 The opening menu for a corrosion database. For any individual choice, associated subchoices appear. Highlight bars are used for actual selection.

$$\text{ETHANE}$$

$$C_2H_6$$

$$CH_3\text{---}CH_3$$

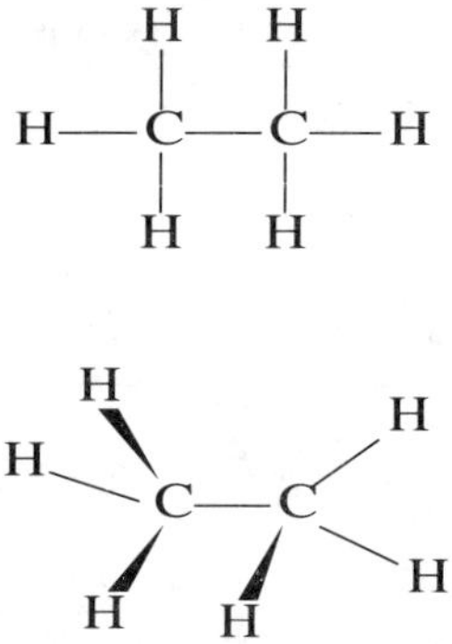

Figure 4.4 The very common chemical ethane can be represented in many ways. All of these are routinely used. Chemical nomenclature systems must be able to handle an even wider variety of systems.

Chemistry is the most advanced discipline in terms of the number, diversity and sophistication of databases. The reason for this progress has been the development of techniques that handle, manipulate and translate different nomenclature systems. Today, access to bibliographic and numerical chemical databases from the smallest PC to the largest online system is provided by sophisticated software written to implement these techniques. Many of these software packages are available to the public and database builders. Some of the more advanced systems have built into them database management capability that can be used immediately upon acquisition.

The most useful key to translation between chemical nomenclature systems is the Chemical Abstracts Registry Number, assigned by the Chemical Abstracts Service (CAS), to each new chemical for which information is abstracted and included in the Chemical Abstracts. This file now numbers over 7 000 000 compounds. Thus, if a particular name, formula, or structure can be associated with a CAS number, then all nomenclature used previously for that number is associated with the compound in question.

Beilstein in the Federal Republic of Germany has developed a similar numbering system for the 1 500 000+ compounds that have been included in their publications. Hopefully, the CAS and Beilstein systems will maintain interconvertibility, which will greatly profit users of chemical information.

The standard chemical database contains:

(i) a software module that handles nomenclature and structural information and provides answers to typical chemistry questions as shown in Figure 4.5

(ii) numerical and other data linked to these compounds through a unique identification number, usually the CAS Registry Number

The features of these software modules vary considerably and reflect the ingenuity of the designers and the projected needs of the users. The recent development of PC packages allows hands-on access for every chemist and provides graphics of publication quality, increasing the ease of manuscript preparation. An excellent introduction to chemical software is provided in the proceedings of a recent symposium on this subject [1].

For engineering materials such as alloys, plastics, polymers, ceramics and composites, the situation is more complicated and less advanced. The problem is exacerbated by the fact that engineering materials rarely have one unique composition and that the material production process drastically affects properties. Thus a designation system for engineering materials must include both typical ranges and actual composition and material processing.

There are many nomenclature systems for engineering materials, and each country has one or two that are widely used. An effort has been started to standardize these systems or at least to provide for the translation from one to another. It will be many years, however, before this will be achieved. The need is clear for a standard system to identify the equivalency of engineering materials. If it is determined that two materials are equivalent, then the data can be combined to form a larger data set.

```
Find all compounds with the following ring pattern...

   (Note: Sometimes the rings may be deeply embedded in a structure)

Find all compounds with the following fragment...

   (Note: The fragment is usually defined as a central atom plus neighbours.
   Matches must be exact. The user might want to include or exclude all
   tautomers.)

Find all compounds with the following specified words or word
fragment...

Find all compounds with the following substructure...

   (Note: For example, all 6-member rings with a sulfur atom ortho- to a nitroso
   group.)
```

Figure 4.5 Typical structure and nomenclature questions that chemist might ask. Large and small scale chemical software can now easily handle these.

In other technical areas, the situation varies. For disciplines where the nomenclature has been systematized at least to some extent, such as geology, or where the objects are easily identified, such as astronomy, computerized designation systems are quite feasible and have been developed. In biology, the situation varies from subfield to subfield. Certainly taxonomy nomenclature is readily amenable to computerization. Gene sequencing is the recent subject of considerable emphasis.

4.2.2 Types of property data

Accessing property data has its own inherent problems. There are three different types of properties that are recorded: intrinsic, test-defined and occurrence.

(i) **Intrinsic** properties are defined by the laws of nature and by the material and its state (temperature, pressure, etc). Intrinsic properties are often expressible as a mathematical relation with a theoretical basis. For example, they can be determined by a variety of experimental procedures and are independent of the measurement technique. Thermodynamic data, density and electron interaction parameters are all examples of these kinds of data.

(ii) **Test-defined** properties are data obtained by following a specific procedure for testing a material. Changes to that procedure will cause changes in the property. In many cases, these tests are formally standardized and the independent variables identified and controlled. The mechanical properties of engineering materials and physiological performance tests are examples.

(iii) **Occurrence** data refers to observations that are made involving counting or enumeration. Medical and bird population statistics are examples.

Each data type has different access patterns related to different numbers of data fields, metadata and associated data. Access patterns are important in determining the physical implementation of a database as well as the suitability of a given database management system for a given data collection. Access to intrinsic property data is usually direct and does not need specification of much associated data. Test data and occurrence data, however, usually require the specification of a great deal of associated data. The contrast is shown in Figures 4.6 and 4.7.

In general, test-defined and occurrence property data have more complicated user interfaces because of the greater number of data fields involved. In these cases, if the search does not yield the desired results, databases often allow users to change one or more specifications and retry the search. In contrast, if a search for intrinsic property data fails, usually the only change needed is to enlarge the data range requested (Figures 4.8 and 4.9).

4.2.3 Searching for property data

Shoshani *et al* [2] have distinguished between two aspects of access patterns or searching: **access type** and **access sequence**. Access type is defined by a single query; access sequence by the relationships between queries.

Search for a property.

    What property: boiling point

    What compound: ethane

    The boiling point of ethane is -88.6 C

Figure 4.6 Search for an intrinsic property. This example is quite straightforward. Other intrinsic properties are function of variables such as temperature and/or pressure.

What property: wear rate

What material: S31600

What test: G-65D

What counterface material: chlorobutyl

What abrasive: sand

What size: AFS/50/70

What feed rate: 231

...

Figure 4.7 Accessing test-defined property data. Because the test results depend on so many variables, many specifications may be required. Of course the user may answer 'all' to any question.

Search for materials with a specified intrinsic property

What property: boiling point

Maximum value: 80.2 C

Minimum value: 79.8 C

No compounds were found with boiling point 79.8 to 80.2 C.

New maximum value: 81.2 C

New minimum value: 78.8 C

3 compounds were found with boiling point 78.8 to 81.2 C. ...

Figure 4.8 Refinement of search for intrinsic property data.

Search for materials with a specified test-defined property

What property: wear rate

What test: G-65D

What counterface material: chlorobutyl

What abrasive: sand

What size: AFS/50/70

What feed rate: 231

...

No materials were found meeting the above specifications.
Which specification do you want to change ? ...

Figure 4.9 Refinement of a search for test-defined property data. If no matches are found, the user can change any one of the specifications and try again.

Table 4.2 Access types for technical databases.

Type	Characteristics
Exact match	Find single value
Range	Find values within a spread
Proximity	Find neighbouring values
Partial	Find values for specified associated data

An access type can be one of four varieties (Table 4.2). An **exact match** is a query that requests a specific data value. A **range search** involves finding data values within a spread or range. A **proximity search** requests data values that are neighbouring to a specified value. A **partial search** looks for records in which one or more of the requested data values are matched, either exactly or within a range. Examples are given in Table 4.3. In practice, different areas of technical data will have quite different access types associated with their use.

Access sequence refers to sequences of queries which are related to one another. Shoshani *et al* [2] also defined two access sequences: local and nonlocal. They define **local access sequence** to mean that the requested data values of the query are close to those of the last query.

Example: Query 1. Compounds melting at 300.0 K
Query 2. Compounds melting at 302.0 K

A **nonlocal access sequence** implies there is no relationship between the requested data values of two successive queries. There can still be a pattern such as:

Query 1. Compounds melting at 300 K
Query 2. Compounds melting at 400 K
Query 3. Compounds melting at 500 K, etc

or it can be totally random:

Query 1. Compounds melting at 300 K
Query 2. Compounds melting at 273 K
Query 3. Compounds melting at 500 K.

Both access type and access sequence should be determined during the planning and design phases of building a database to allow accurate definition of data relationships and optimal database performance. If patterns of access can be identified for certain data fields, the database can be set up to use these items primarily as access paths and thus greatly reduce the demands for random access.

Several considerations must be taken into account when searching for data which are independent variables. The first consideration is to make the user aware of what

Table 4.3 Examples of access types.

Exact Match	Compounds that melt at 300.0 K
Range	Compounds that melt between 300.0 K to 310.0 K
Proximity	Compounds that melt at 300.0 K $\pm$ 0.5 K
Partial	Compounds that melt at 300.0 K and are either yellow or soluble in water
	(A compound is retrieved if it is **either** yellow **or** soluble in water **or both**)

data values exist. One method is to index each independent variable data value so the user can choose from those included in the database (Figure 4.10). In some situations, too many different data values may exist, precluding easy indexing or display.

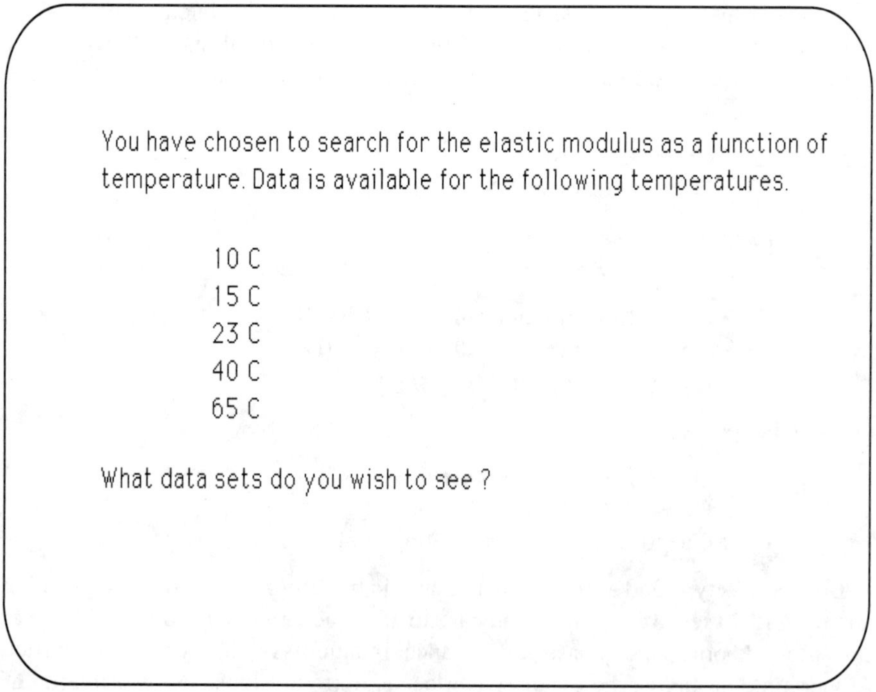

Figure 4.10 Searching for property data that are functions of independent variables or test parameters. For test-defined data, the number of parameters might be quite high and not easily displayed.

4.2.4 Inexact data

Handling searches where the requested data are not exactly matched by data in the database is a major consideration as illustrated by the following three requests for the data shown in Figure 4.11.

What is the vapour pressure at 300 K?
What is the vapour pressure at 305.5 K?
What is the vapour pressure for temperatures between 295 K and 325 K?

The first request is handled simply since an exact match can be made to the temperature value of 300 K.

The second request is more difficult because no value in the database exists for the requested temperature. However, enough data exist that a reasonable estimate can be made, either automatically by the database or by the user if shown enough data values. For example, the two bracketing values could be given.

At $t=300, vp = 20.8$
$t=310, vp = 30.8$

Providing just two values would restrict interpolation to linear interpolation which is rarely the most accurate. Making use of more neighbouring points can improve the interpolation considerably. Therefore, displaying the four surrounding points would be useful.

At $t=290, vp = 14.4$
$t=300, vp = 20.8$
$t=310, vp = 30.8$
$t=320, vp = 49.4$

An even more effective alternative would be to store an equation that fits the data. When a query is made requesting a value that requires interpolation, a procedure is

Temperature, K	Vapour pressure
250	10.0
260	11.0
270	12.0
280	13.2
290	14.4
300	20.8
310	30.8
320	49.4

Figure 4.11 Sample vapour pressure data.

invoked to calculate the result, not leaving it up to the user. Some technical databases already provide this capability.

The third query is more of a problem, first, because it is a range search and, second, because it requires extrapolations outside the table of values. Range searches are discussed in the next section. For extrapolation, polynomial curve fits should never be used because they diverge rapidly. It is best to use an extrapolation formula if one is known. Otherwise, linear extrapolation is safest, or it may be better simply to report that no data are available.

Other alternatives exist for handling the last query. One procedure would be to give the requester the full vapour pressure–temperature table suitable for downloading so further analysis could be completed under direct user control. Alternatively, an expert system interface could be built to guide users to the correct manipulation of the data. Virtually no work has been reported on these approaches.

There are no set procedures for handling the above problems, but some approaches have been developed. During analysis of the user requirements, the independent variables for queries of the nature just discussed can be identified. The user interface should be written to intercept these queries and invoke interpolation or data-fitting solutions when they are needed. This extends the database software because the user interface will operate beyond the immediate context of the database management system. Until recently, few DBMS's allowed the execution of formulae such as are found in much simpler spread sheets.

4.2.5 *Multi-dimensional searching*

More complex search requests may easily be formulated, for example, a search to find materials that have data on two separate properties, X and Y. A database might contain such data but in different data records. The request becomes a database-wide search, and the system must recognize that even though property X is in one record and property Y in another for the same material, that material meets the query specifications. This request is not easy for many DBMSs to handle.

As the size of technical databases grows and networks and systems integrate individual databases together, requests similar to the last example will have to be performed across databases, a formidable challenge.

4.3   DATA TRANSFER AND DOWNLOADING

Data retrieved from a technical database often need to be transferred to another computer or to other software. This process is called **data exchange** or **data interchange**. Sometimes, the form of the interchange is well-specified by a data interchange format. When transfer takes place from a database on a larger computer to a personal computer, the process is called **downloading**. In this section, we discuss both processes, beginning with downloading.

4.3.1 Downloading

Downloading means transfer of search results from a database (usually on a mainframe) to a locally controlled computer for further processing, often into an application database (Section 3.6). This could involve simple reformatting to make the output more attractive, transforming the data for graphical display, or saving the data for further use. The concept was developed when databases, especially bibliographic databases, were only available on large, costly systems that provided little post-retrieval capability. Until recently data system vendors were apprehensive about downloading because of the potential for reduction in system sales. In addition, fears were raised about the transfer of entire databases. However, the inevitability of downloading, especially with the spread of PCs, has been recognized, and the approach now is to allow some downloading with recovery of lost revenue by increasing charges.

A second reason for downloading has arisen as institutional databases have emerged. In these cases, users are encouraged to access a central database and transfer data to their PCs for their own work. Often the use of these central technical databases becomes routine, and the same type of search is performed repeatedly. Online data networks also anticipate this pattern of use.

If a database is to support downloading, provision must be made to put data into an appropriate format. Because it is impossible to anticipate every use, databases should allow users to define different formats, equivalent to 'views' or 'schemas' in some database systems (see Chapter 8). When the number of data fields becomes large, specification of the format can be tedious, complicated and difficult to remember. Data records with 50 fields that are downloaded for both immediate and future use will need good documentation for future interpretation. The format could be self-defining as follows:

```
field name = field value
```

For example,

```
chemical = water,
formula = H₂O
molecular weight = 18.0154
melting point = 0
units = °C
boiling point = 100
units = °C
```

For large data sets having many measurements repeated for different values of an independent variable, this format can consume much extra storage. An alternative technique would be to include a defining header field that gives the order:

```
chemical, formula, molecular weight, temperature units,
melting point, boiling point, water, H₂O,18.0154,°C,0,100,
carbon dioxide, CO₂, 40.216, ...
```

A third possibility, close to an object-oriented approach (see Section 9.6), is to provide at the start of the file a simple high-level language program which can read the file and print it again in the correct format.

Database builders, distributors and users must recognize that downloaded files are often preserved for future use, and years later someone else should be able to identify their format easily. Many times the downloading of data becomes routine or automatic. For example, experimental data may be routinely put into a database and then passed to data programs. The commands to perform downloading can then be captured and executed each time. Downloading into other programs is a common occurrence and will be discussed more fully.

4.3.2 Data exchange formats and standards

As the number of databases grows and application software proliferates, neutral data exchange formats may be needed so that all software for a given type of application may use the same downloading format. In certain technical areas, standard data exchange formats are being established, both formally and informally. The formats provide for exchange of large amounts of data, such as data entry to a database by different groups or for the exchange of large databases in their entirety. In other situations, the format will permit easy transfer from a database to software.

The purpose of a neutral format is to reduce the data exchange problem considerably. Data transfer simply has to be implemented between a database and the neutral format and back again. Transferring data from one database to another is accomplished by first translating from the database storage scheme to the neutral format, then from the neutral format to the second database storage scheme. The addition of a new database storage scheme to the transfer process does not create any work aside from one translation into and one out of the format.

One of the first such database exchange standards for technical data, the highly successful EXFOR (Exchange Format) [3], was developed for neutron scattering data under the leadership of the Nuclear Data Section of the International Atomic Energy Agency, working with neutron data specialists in Europe, the United States, the Soviet Union, and Japan. Groups in other technical data areas are very active also, as summarized in Table 4.4.

Some important general points can be made related to data exchange standards for technical databases.

(i) Data types, not data occurrences, should be defined. Often data values are mixed up with data field names. For example, the exchange format should provide for data fields, e.g., 'colour,' and not have multiple fields called 'red', 'yellow' and 'blue'.

(ii) The formats should be easily extended for both new data fields and different data values. Fixed field formats do not have the flexibility of variable field formats. Variable field formats also allow data compression by eliminating the need for part empty fields (see Section 7.2).

Table 4.4 Some activities on technical data exchange formats.

Subject	Groups Involved
Surface Science	VAMAS, ASTM Committee E42 [4,5] (see Figure 3.2)
Material Properties	ASTM Committee E49, CEC [6,7]
Crystallographic	NIST (formerly NBS), Cambridge Crystallographic Centre, International Centre for Diffraction Data [8]
Protein Structure	Brookhaven National Laboratory, Los Alamos National Laboratory, National Library of Medicine [9]
Geophysical Data	World Data Center [10]

(iii) The computer can easily determine and translate complex structures. If it is not necessary to preserve human readability in an interchange format, complicated structures can be established and interpreted by computers. Some groups distribute software that reads and writes into and out of their format.

(iv) New formats should not ignore past work. Many different formats from technical areas exist. These should be examined before a new format is defined. Several simplistic and naive formats have very heavy use, which demonstrates that a simple format is often sufficient.

It is interesting to note that chemical databases, the most advanced technical database area, have virtually no data exchange standards.

4.3.3 Use of many databases

With the advent of the PC generation of DBMSs, a virtual explosion of technical databases has occurred. Most applications requiring the use of technical data rarely find that one data source can provide all needed information. Users need access to many data sources to solve their problems, using two or more databases if necessary. One of the first actions users will take is to compare data from several different sources.

The proliferation of databases results not only in a diversity of DBMSs but also in a diversity of terminology. The comparison of data from different databases must therefore overcome two problems:

(i) different database formats, schemas, physical representations
(ii) different terminology for the same information.

The first of these has been treated in the above discussion of neutral formats. Here we will concentrate on the second, the problem of terminology, which is most keenly seen in online systems, database networks and CD-ROMs. All of these can contain many databases built by different groups using different terminology. The two diverse approaches to harmonization of terminology are:

(i) to establish rigid standards and require every database to conform or

(ii) to develop software that will translate every term into all of its synonyms.

Neither approach is realistic, and no standard will be universally accepted since the cost of building an all-encompassing data thesaurus is astronomical. A combination of standards and a data thesaurus will be the pattern, with the actual combination of these two determined by how quickly the collections of databases become important. There is every indication that this will be soon.

Sometimes a dominant group will establish *de facto* standards by building a series of compatible databases. However, it remains to be seen whether these efforts can in fact stand by themselves or will eventually need to be integrated into larger systems.

4.4 DATA DISPLAY

Users will want to see data displayed in approximately the same way that they are accustomed to seeing the data on the printed page. Graphs, tables and text are all expected. Databases will define one set of output options, but users will also want to create their own displays.

4.4.1 Graphics

Scientists and engineers have become accustomed to graphical displays of technical data for reasons summarized in Table 4.5.

Personal computers have made graphics accessible to the same users with no great effort, but for other users graphics can still be a major problem, for the following reasons.

(i) Users often do not have a full graphics terminal.

(ii) DBMSs do not routinely create files directly usable as input to graphics packages.

(iii) The multiplicity of graphics packages, database management systems and terminal types presents virtually unlimited combinations, causing great confusion.

Table 4.5 Advantages of graphical display of technical data

Compress large amounts of data
Provide comparison between different results
Show trends and analysis
Can be appealing and attention orienting

(iv) Communications speed between central computers and terminals is usually too slow for heavy graphics production by online systems.

Some technical databases have associated graphics packages that easily permit the transfer of data between database and graphics. The display of diagrams, especially chemical structures, can now be accomplished by extremely powerful packages that provide such a wide variety of choices that every chemical database builder can assume an appropriate package exists.

Other diagrams are sometimes associated with technical data, but these rarely can be drawn as part of normal database usage. The major exceptions to this are geometrical drawings associated with engineering or electrical circuitry design. Techniques for handling these design databases or geometrical diagram databases are beyond the scope of this book.

In using technical databases, users will want to graph data from different data records or even from different databases. Data conversion must be possible in an easy, straightforward manner. Units must be made uniform, ranges of independent and dependent variables adjusted, and ancillary information brought together and displayed in a consistent manner. The full potential of graphical displays for technical databases can only be realized when users are able to make graphs without bother or fuss. That day is almost here.

4.4.2 Tables

For technical databases, the capability to create tables today lags behind graphics, which is surprising considering that spread sheet software has gained such wide acceptance. Most typical technical data tables can easily be handled by spread sheets, yet virtually no technical database has that kind of power. Users arbitrarily want to define rows and columns, to move rows and columns around, to create headings spanning an arbitrary number of columns, and to be able to move (scroll) through rows and columns too large to fit on the screen all at once. These needs are simple but not satisfied. A technical database builder should examine spread sheet displays and try (while complying with copyright rules) to incorporate features into the display of tables. An alternative is to download the tables to a PC and use a good software package on the PC to display the tables.

4.4.3 Text

The display of textual information for technical databases has not been well studied. The subject has importance for at least three types of systems: (1) online journal databases, (2) expert systems and (3) full-text databases. Online journal databases and expert systems (Chapter 12) are outside the scope of this section. Full-text databases contain the equivalent of a handbook with large amounts of explanatory text (Figure 3.3). In this case, data are presented in the text as well as in tables and graphs. Some attempts have been made to build such databases, ironically without the tables or graphs. These databases are searched, using conventional text-handling software. Virtually no attempts have been made to approach the problem from the viewpoint of finding numeric data. One exception is work by Chemical Abstracts to extract numeric data from their online journal databases [11].

4.5 PUBLICATION

One common output from a database system is a printed page; good printing is an essential part of the user interface. It is true that interactions with the database are usually through a display screen. However, once a user finds the data which are actually needed, they are normally printed. Even when a user retrieves a file from the database and transmits it to a PC or workstation, usually all or part of it is printed at some later time. Most printing, however, is direct from the database management system to the printer in response to a user request. Occasionally print is used to publish part of a database.

The reason that published data in a printed form are so important is that people almost invariably prefer to read from a printed page rather than from a display. This is unlikely to change in the near future, so good-quality print is vitally important as an end product.

4.5.1 Typesetting

The subject of printing and desktop publishing has recently been reviewed in detail by Kleper [12]. Here we discuss only the aspects of the typesetting process which are important for databases: composition, fonts, graphics and mathematics.

The first two items concern any database system; the last two are particularly important in science and engineering. They are briefly discussed below, and a few comments on printers follow.

1. Composition

Composition concerns the layout of type, graphs, images, etc on a page. It includes indentation, line spacing, margins and correct alignment of text and tables. Unfortunately, databases usually at most provide only limited facilities for composition, those necessary to print tables. To obtain attractive printouts, files of data must be transferred from the database to a personal computer or workstation and one of many available desktop publishing systems used to improve their appearance.

2. Fonts

Technical data require not only the three common forms of fonts—normal, bold and italic—but also mathematical symbols, Greek characters, superscripts and sub-scripts. These are needed in a wide range of sizes. Unfortunately, it is unlikely that a technical database system can cope with these requirements, but a good word-processing system should provide what is needed after files have been transferred from the database systems.

3. Graphs

Graphical output is essential in science and engineering, and a good database system should provide facilities for the drawing of high- quality graphs (see Section 4.5.2). However, the capability to integrate these graphs onto the printed pages of a report is unlikely to be available. A good word-processing or page-description package should make this possible.

4. Mathematics

The printing of mathematical formulae in the headings of tables or figures or in a report from a database system is often needed, but all too often is not available. Typesetting systems like TeX [13] permit the printing of complex mathematical formulae, but they are not easy to learn and are not designed to handle data from a database system. Again, a good word-processing system may be sufficient.

4.5.2 Printers

The days of metal type are almost behind us, although metal-type printers still give the highest quality print at a reasonable price, if only for a limited number of fonts. Almost all modern printers are now forms of dot matrix printers and can be classified into three main groups: pin-dot matrix, laser and ink-jet.

All of these construct the characters or graphs on a page as a collection of dots, overlapping and so small and so close together that they deceive the eye. The collection of dots representing each font is either stored or generated as needed. The range of fonts can be wide and dots can also be used to draw graphs or produce images. Thus, dot printers are a great improvement on earlier metal-type printers. However, to obtain the same quality of type, a dot printer needs to be able to print over 1000 dots per inch, now possible on only the most expensive phototypesetters. The usual compromise is 300 dots per inch, which still looks good, if not perfect.

1. Pin-dot matrix

The lowest quality of these are the 9-pin-dot printers, which produce poor but acceptable type at very low cost. With improvements in technology and because of user demand for better quality, for multiple type sizes and fonts, and for clear graphs, the 9-pin-dot printers are being replaced by slightly more expensive 24-pin-dot printers. These have a much improved quality, being able to print 300 dots per inch.

2. Laser

Although 24-pin-dot printers claim to match the 300 dot per inch resolution of the laser printers, the quality is not as good. In a laser printer [14] a laser is focused on the surface of a drum, creating a difference in electrical charge which attracts toner powder that is then fused to the paper by heat and pressure. The toner powder used by the laser printers gives greater contrast than the standard printer ribbon used by the pin-dot printer.

The laser printers avoid the mechanical impact of the steel pins and are thus quieter and faster. Although laser printers give higher quality, they are more expensive. Often one laser printer, because of its cost, is shared among many users.

3. Ink-jet

Newer ink-jet printers [15] fall between the laser and pin-dot printers in price and quality. The development of this new technology, which simultaneously fires tiny drops of ink from up to 50 nozzles to produce a resolution of 300 dots/inch, may eventually replace the pin-dot and laser printers. These printers also have the capability to print in colour.

Clearly the printer chosen depends on its use. A database report being sent to a customer should use the best-quality printer available—probably a laser printer. However, it does not make sense to tie up an expensive printer on routine dumps from your database.

4.6   DATA EXHIBITS

Often databases contain data records corresponding to published tables, graphs and paragraphs which are called data exhibits. Exhibits often are compact displays allowing large amounts of data to be shown, either for completeness or for contrast. When databases contain data equivalent to published exhibits, their reproduction is important for many reasons: data might be meaningful only in the context of the full exhibit; the database might be used to support the publication process and to produce high-quality publishable output; users might be accustomed to seeing the data in a certain format (as published) and feel most comfortable with a similar computer display. Engineering databases within companies frequently make exhibits, i.e., an engineering drawing, the primary entity for a database to preserve an established engineering practice. In some cases, individual data fields can be searched; in others, only complete exhibits. Normally, all data are displayed only in an exhibit format.

Another common reason for exhibits in databases is user reinforcement for data integrity purposes. Engineers often use standard data sources, either proprietary or public, and have grown accustomed to seeing these sources in a particular format. When being weaned to computerized databases covering the same information, many users feel most comfortable seeing the familiar form. Once familiarity with computerized data is established, these users can be moved to more appropriate database displays. Exhibits are also useful in validating data, after data entry or for user reassurance.

4.7 SUMMARY

Technical databases are subject to a wide variety of uses and must support the uses requested. In planning a technical database, these uses must be identified, articulated and supported. Since the emphasis in each technical area differs, the general classification outlined here remains useful in this planning work.

REFERENCES

[1]   *Graphics for Chemical Structure*, Ed. W. Warr, *American Chemical Society Symposium Series No. 341*, American Chemical Society, Washington, DC, 1987.

[2]   A. Shoshani, F. Olken and H. Wong, Data Management Perspective of Scientific Data, in *The Role of Data in Scientific Progress, Proceed 9th CODATA Conference*, Ed. P. S. Glaeser, North-Holland, Amsterdam, 1985.

[3]   Detailed Documentation of Format and Compilation Rules: *EXFOR Manual*, IAEA-NDS-3, Revision 85/8, International Atomic Energy Agency, Nuclear Data Section, P.O. Box 100, A1400 Vienna, Austria, 1985.

[4]   W. A. Dench, L. B. Hazell, and M. P. Seah, VAMAS Surface Chemical Analysis Standard Data Transfer Format with Skeleton Decoding Programs, *NPL Report DMA(A)164*, National Physical Laboratory, Teddington, Middlesex, UK TW11 0LW, July, 1988.

[5]   ASTM Committee E42 on Surface Analysis. Contact ASTM, 1916 Race Street, Philadelphia, PA 19103.

[6]   ASTM Committee E49 on Computerization of Material Property Data. Contact ASTM, 1916 Race Street, Philadelphia, PA 19103.

[7]   H. Kroeckel and G. Steven, The Integration of Materials Data Banks into a European Information Service Network, in *The Role of Data in Scientific Progress, Proceedings of the Ninth International CODATA Conference*, Ed. P. S. Glaeser, North Holland, Amsterdam, 1985.

[8]   A. D. Mighell, C. R. Hubbard, and J. K. Stalick, NBS*AIDS80: A FORTRAN Program for Crystallographic Data Evaluation, *NBS Technical Note* 1141, U.S. Department of Commerce, National Institute of Standards and Technology, (formerly National Bureau of Standards), Washington, DC, 1981.

[9]   For Protein Structures: Protein Data Bank: Atomic Coordinates and Bibliographic Entry Format Description, available from the Protein Data Bank, Chemistry Department, Brookhaven National Laboratory, Upton, NY, 11973.
      For DNA Sequencing: contact GenBank, Los Alamos National Laboratory, MS K710, Group T10, Los Alamos, NM 87545.
      For Protein Sequencing: D. G. George, H. W. Mewes and K. Kihara, A Standardized Format for Sequence Data Exchange, Protein Sequence and Data Analysis, 1, 27–39, 1987.

[10] Contact: The World Data Center for Solid Earth Geophysics, 325 Broadway, Boulder, CO 80303.

[11] S. W. Terrant et al., Online Searching: Full Text of American Chemical Society Primary Journals, *J. Chem. Info. Comp. Sci.*, 24, 230–35, 1984.

[12] M. L. Kleper, *The Illustrated Handbook of Desktop Publishing and Typesetting*, Tab Books Inc., Blue Ridge Summit, Pennsylvania, USA, 1987.

[13] D. E. Knuth, *TeX and Metafont*, Digital Press, New York, 1979.

[14] J. Parris, Non-impact Printing, *Computer Systems*, 6, No. 10, 51–3, 1986.

[15] J. D. Rhodes, Managing the Development of the HP Deskjet Printer, *Hewlett-Packard Journal*, 39, No. 5, 51–4, 1988.

Chapter 5

User Interfaces

5.1 INTRODUCTION

Unless a database system is intended solely for use by the builder, other users will access the system through the user interface, and to the user the system is the interface. Over the last 10 years, the importance of the interface, especially to the casual or occasional user, has been recognized and the principles of a good interface understood. Many technical database builders, only now understanding that their success depends critically on the quality of the interface, are beginning to give attention to this aspect.

The purpose of an interface is to allow the user to select a particular action, to execute it, and to view the result. Consequently, the basic principle of good interface design can be expressed simply:

(i) all avenues of action must be clearly and unambiguously presented to the user;

(ii) all results must be displayed to the level of completeness desired by the user.

In both cases, it is the **user** who defines what is needed, not the interface designer. As the details involved in interfaces are discussed, this point must always be kept in mind. Cleverness, dexterity and virtuosity can all be carried too far, thus confusing users. At the other extreme, terseness and sparseness perplex and leave the user uncertain as to what to do next. In this chapter we draw on our own experience and the research of others to outline the major considerations needed to achieve harmony between the two extremes.

5.2 COMPONENTS OF AN INTERFACE

The purpose of the interface is to allow the user to navigate through a sequence of possible events and actions to find the desired information. Therefore, at any one time, the user must be able to:

1. See the results of the last action
2. Know the next primary actions available
3. Know the next secondary actions available, if any
4. Find help if needed
5. Understand the present status of the system

Given the limitations on typical computer terminals, often all this information cannot be displayed at once. However, details on how to access all the information must be available.

1. Seeing the results of the last action

The first thing a user wants to see is what happened as a result of the last requested action. Was the file erased? What materials were found with the specified properties? How many chemicals have their melting points within a certain range? If the action was simple or merely procedural, the presentation can be very terse (Figure 5.1).

Often the action results in too much information to be conveniently displayed on one computer screen. Then the interface designer must resort to a sequence of displays whose logic must be apparent to every user. Figures 5.2 and 5.3 give one example of a two-tiered display with the first screen telling how many data sets meeting the user specification were found and the second screen showing the full data set for one hit.

```
                    Search by Database


You have selected access by search of databases.  The
following  databases are accessible on the MPD Network at
this time:


        1 - MIL-HDBK-5 Metallic Materials Design Handbook
        2 - Aerospace Structural Metals Handbook (ASMH)
        3 - Toughness of Steels (STEELTUF)
        4 - Toughness of Marine Steels (MARTUF)


M - Main Menu     P - Previous screen     X - eXit   ? - Help


Enter selection:  3
```

Figure 5.1 Screen from MPD Network System for selection of databases.

Matches from unknown spectral lines search (total 5)
    data   = binding energy
    line energy   = 540.00 eV
    energy tolerance = ± 1.00 eV

Result   Line   Phys.
No. Energy El Designation Compound   State

 1   539.6 Sb   3d3/2   Sb2O3
 2   539.8 Sb   3d3/2   Sb2O4,   beta
 3   540.2 Sb   3d3/2   Sb2O5
 4   540.2 Sb   3d3/2   USb3O10
 5   540.2 Sb   3d3/2   USbO5

PageDown/PageUp   Main Menu: 0   Display: (Result No. 1–5) >

Figure 5.2 A screen summarizing the results of a search of the NIST X-Ray Photoelectron Database. Full data for any entry can be displayed by entering the Result No. The full display is shown in Figure 5.3.

Significant software and hardware improvements now available on many personal computers have gone a long way towards providing innovative methods for displaying results. For example, windows, which 'pop-up' and contrast vividly with the background (Figure 5.4), allow users the option of displaying information on one part of the computer screen while leaving other information in place. Scrolling left–right or up–down also adds flexibility. Cursor controllers and mouses allow direct access to a particular location on a screen [5]. All these features reduce the size limitations of the computer terminal screen and allow creative display of results.

2. Knowing the next primary action

Unless the user has performed the final logoff from a system and is about to walk away or switch off the computer, there is always a **next** step. Nothing is more

```
 NIST XPS Database


 Element: Sb  51


 Compound Name:     Sb2O3
 Compound Type:     Oxide,Sb+3
 Empirical Formula:  O3Sb2
 Physical state:


 Data Type: photoelectron line      Binding Energy: 539.6 eV
 Line Designation: 3d3/2

                                     Data Quality: reliable, corrected value
 Calibration:
 Charge Reference:  AC
 Reference:          DBLG83
 Complete Reference:
 DELOBEL, R., BAUSSART, H., LEROY, J., GRIMBLOT, J., GENGEMBRE, L.,
 J. CHEM. SOC. FARADAY TRANS. I 79, 879 (1983)


 Previous menu: 1      Main menu: 0      Please enter choice >
```

Figure 5.3 A full display of data for an entry in the summary display from Figure 5.2.

frustrating than not knowing what to do. One classic example is in exiting from BASIC on a PC using Microsoft DOS (TM). The command SYSTEM is cryptic, non-informative and virtually unguessable as compared to words such as END, BYE, EXIT, or QUIT.

The primary actions available to users must be clearly visible or quickly obtainable. The common methods for making the choices known are menus and commands, both of which will be discussed later in this chapter. For novice or casual users, menus are attractive because the choices can be presented all at once, ordered

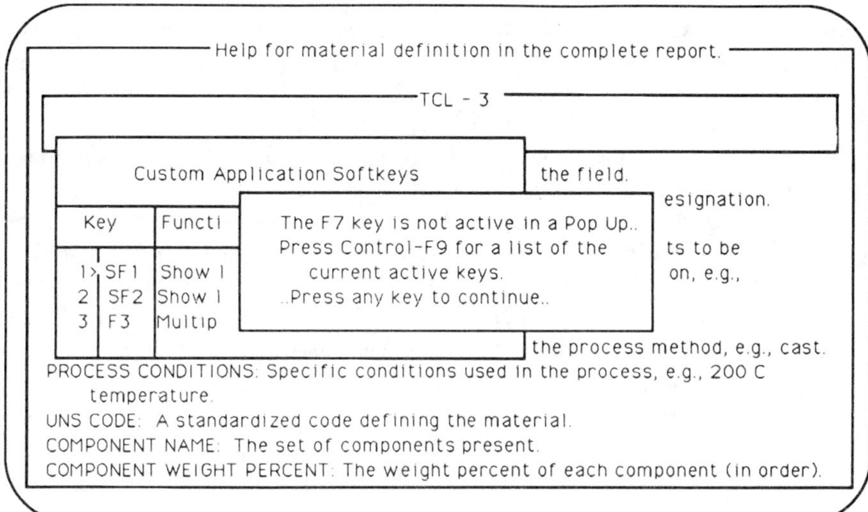

Figure 5.4 Example of windows. Here a mistake is emphasized by contrasting the information to that already on the screen.

or highlighted according to importance or by probability of selection. The major difficulty with menus arises when the number of actions is so large that they do not fit easily on one screen. A hierarchical series of menus can be used; however, when users become proficient they are irritated by the time taken going through many menus to execute some action. A method of executing a complex action in one step should therefore be provided to the frequent user.

For the novice, menus must be very simple. The syntax and semantics of the commands, namely what to do and how to express the action, can be insurmountable barriers that stop use before it has even started. This is especially true when the options are many and the system 'help' features are poor.

Well-designed systems can feature both menus and commands and allow for switching from one to another. Another possibility is gradual transition from menus to commands as user proficiency grows. This may be done either in a single session or from session to session.

Ambiguity of actions must be avoided to prevent user alienation. With menus, ambiguity usually can be avoided because the descriptive text that goes with any selection can be explanatory. For commands, ambiguity is a major problem, especially when the list of command options is not given or when the commands are shortened or abbreviated.

3. Knowing what secondary actions are possible

Often actions displayed in menus represent only a fraction of the total possibilities. The number and variety of the options may be too large to fit on a screen presenting primary action choices. The availability of secondary actions must be made known. They can be summarized at the bottom of a screen (Figure 5.5) or made available by windows or secondary menus (Figure 5.6). Another possibility is plastic cards that fit around function keys on the keyboard. These are popular for word-processing and spread-sheet packages. In any case, the secondary actions need not be as prominent as primary choices but still must be easily accessible.

For technical databases, unsupported search strategies and poorly formulated queries may require users to undo the last few steps of a search rather than start over completely. Consider the example in Figures 5.7 and 5.8. The sequence of screens in Figure 5.7 establishes the search, and the results are displayed in Figure 5.8. However, too many plastics have properties that satisfy the search request, and the user will realize that the whole range of plastics should first be narrowed to a particular class. A path that returns to the question 'Do you want to select a class of plastics?' (screen 2 of Figure 5.7) is needed.

```
Page 3 of 4


                    COMPLETE REPORT OF RECORD


        MATERIAL DEFINITION:
        Class:                  METAL
        Subclass:               FERROUS
        Common Name:            STEEL
        Principal Component:    Fe
        Second Component:       Cr
        Grade:                  52100
        SPECIFICATION:
        Form:                   BAR
        Process Treatment:      HT
        Process Conditions:     350F TEMPER
        UNS:
        Component Name:         C,Cr,Mn,Si
        Component Weight %:     1.0,1.5,.5,.3




        Use <PAGE UP> and <PAGE DOWN> to move around.
        Use <F1> to see definitions for items on this page.
        Use <ESC> to return to the summary screen.
```

Figure 5.5 Secondary information displayed at the bottom of a screen.

SELECTION SCREEN #2: TRIBOLOGICAL DATA AND CONDITIONS

TRIBOLOGICAL DATA:
Wear Coefficient
Wear Rate

TRIBOLOGICAL CONDITIONS:
Wear type
Contact Geometry
Motion Type
Load
Contact Pressure
Temperature
Velocity
Distance
Counterface Material
Contact Enviroment
Standard Test

Custom Application Softkeys

	Key	Function
1 >	SF1	Show list of attached files
2	SF2	Show list of volumes
3	F3	Multiple line TCL

Figure 5.6 Secondary information displayed as a window.

4. *Finding help when needed*

Help, that is, advice, aid and guidance to users, is important and takes three basic forms:

   (i) What actions are possible?
  (ii) What went wrong and how can we fix it?
(iii) What does something mean?

The first form can be used to explain actions in detail and differentiate between similar actions. The messages should be short and to the point because users will not want to read many screens of text. Popular word-processing and spread-sheet programs have successfully used help messages to achieve widespread acceptance among non-computer-literate users. The same positive impact is expected by technical database users.

The second form of help message explains what went wrong and how to correct mistakes. Computer professionals raised on the cryptic error messages used by large computer manufacturers might view this form of help as pampering. Personal computers, however, have changed users' expectations, and responsive software producers now fully explain errors and clarify corrective actions. The ultimate

```
Select a material class

 metals - 1
 ceramics - 2
 plastics - 3

Enter Choice >  3
```

```
You have selected "plastics"
as your materials class

Do you want to select a class
of plastics - y or n

Enter Choice >  n
```

```
Now, select a property

density - 1
composition - 2
strength - 3

Enter Choice >  3
```

Figure 5.7 Typical search screens for finding a property of a material. Results are shown in Figure 5.8

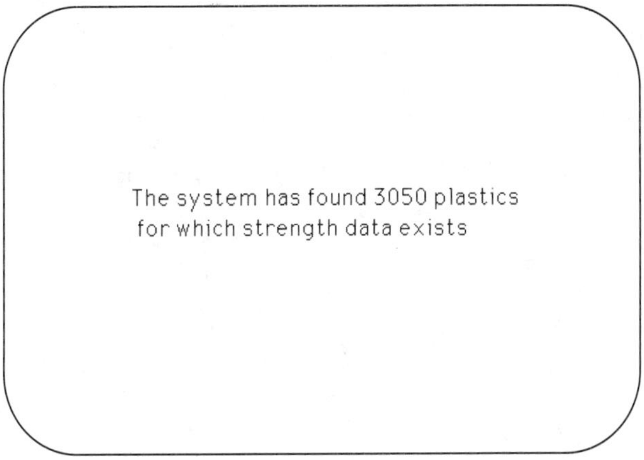

Figure 5.8 Results from a typical search for finding a property of a material. The number of results is too large too handle.

strategy, of course, is to find mistakes before they are made. For example, before a file is deleted, the user must verify that deletion is indeed desired. Mistakes, however, will always be made and will need to be corrected.

The final form of help message relates to what the results mean. This can be as simple as the term definition, i.e., FTU means 'ultimate tensile strength in units of MPa'. More complicated interpretations may also have to be provided.

In particular, the user interface needs to provide help with the less common terms or data fields that may not be familiar, even to competent users. Some database systems allow the typing of a question mark followed by a search term to display a thesaurus record for the term (Figure 5.9). The amount of information provided does not swamp the user, and only a few paragraphs are available in a typical thesaurus entry. However, note that the building of a data thesaurus is very expensive and often open-ended.

A common problem for technical database builders is how much to assume their users know. To preclude misinterpretation or misuse, it has been suggested that, for certain systems, users be given a test before proceeding. However, technical databases will be used for everyday scientific and engineering work by as wide a variety of people as now use published data sources. It has to be assumed that users will exert the same professional judgment on data from databases as they do for data from publications.

5. *Learning the system status*

The last component of the user interface is a facility for users to know where they are in the system and what the system is doing when a long computation is underway. The user is more comfortable with a 'system searching' message than a blank screen. This search can take the form of a series of dots placed on the screen

```
The Term UTS is TENSILE ULTIMATE STRENGTH


STANDARD TERM:  tensile ultimate strength
DESCRIPTION:     The maximum tensile stress which a material is capable of
                 sustaining. Tensile strength is calculated from the maximum load
                 during a tension test carried to rupture and the original cross-
                 sectional area of the specimen.  Values designated Ftu are
                 statistically based design    (A or B) values, or specification
                 (S) values.


LABEL = tens. ult. str
BROADER TERM:  indep/prop variables, tensile properties.
STANDARD UNITS:  pascal.
USED FOR:  ultimate strength, Ftu,T, ultimate tensile strength, tensile  ultimate
           strength, tensile strength, tensile ultimate, tenacity, U.S., U.T.S.,
           maximum tensile stress, Ftu, F{sub tu}, F{sub tu}, T.U.S., UTS,
           ultimate strength, tensile strength, tensile ultimate, tenacity, U.S.,
           U.T.S., maximum tensile stress, Ftu, F{sub tu}, F{sub tu}, T.U.S.,
           UTS, Ftu,max, Ftu,min, Ftu,RT,init.


:Enter your choice of Property(s):
```

Figure 5.9 Thesaurus help from the MPD Network System.

after every 5 or 10 items found. Or the number of searches completed can be displayed after every 5 to 10 searches. A status line at the bottom of a screen can display which system options have been activated.

5.3   THE DESIGN OF GOOD USER INTERFACES

We have summarized five components of the interface that are needed for the user to obtain the most from a database. When combined with the philosophy of screen design as given in this section, good user interfaces will result. Good interfaces will have: consistency, clarity, crispness, accuracy and a logical layout.

1. Consistency

The user interface must be consistent throughout a session and from one session to another. Commands must maintain their meaning. Abbreviations or mnemonics must not be redefined. The same type of information should always appear in the same location.

Consistency is important to facilitate user actions and to minimize wrong choices because of user eagerness or misinterpretation. Figures 5.10 to 5.12 show six screens from a materials information system that maintain consistency with respect to presentation of information, choices and responses.

```
          Search by Database                                    SEARCH

 You have selected access by search of databases.  The following databases are
 accessible on the MPD Network at this time:


        1 - MIL-HDBK-5 Metallic Materials Design Handbook
        2 - Aerospace Structural Metals Handbook (ASMH)
        3 - Toughness of Steels (STEELTUF)
        4 - Toughness of Marine Steels (MARTUF)


 M - Main Menu              P - Previous screen     X - eXit    ? Help

:Enter selection:   3
```

```
           STEELTUF Database                                    SEARCH

   You have selected the Toughness of Pressure Vessel Steels (STEELTUF)
 database.

   Characterization of Database:

 TYPE:               Individual test results from producers and users
                     of steels considered for pressure vessels
 TEST METHODS:       not documented online; traceable in some cases with
                     effort
 TRACEABILITY:       Individual data online; detailed documentation not
                     available
 EVALUATION:         EPRI/MPC/MRCS overview; not formally evaluated
 UPDATED:            June, 1985; stable database, no further updating
                     likely
 ORIGINAL SOURCE:    Survey by EPRI/MPC; assembled by MRCS/Oldfield
 CONTACT:            Dr. Martin Prager, Acting Executive Director,
                     MPC 345 E 47th St, NYC;  Telephone 212-705-7697
 ONLINE SOURCE:      MPD NETWORK; Telephone 614-447-3706

 M - Main Menu       P - Previous screen     X - eXit    ? - Help

:Press RETURN to continue:
```

Figure 5.10 Two of a sequence of six screens from the MPD Network that illustrates consistency of location and style.

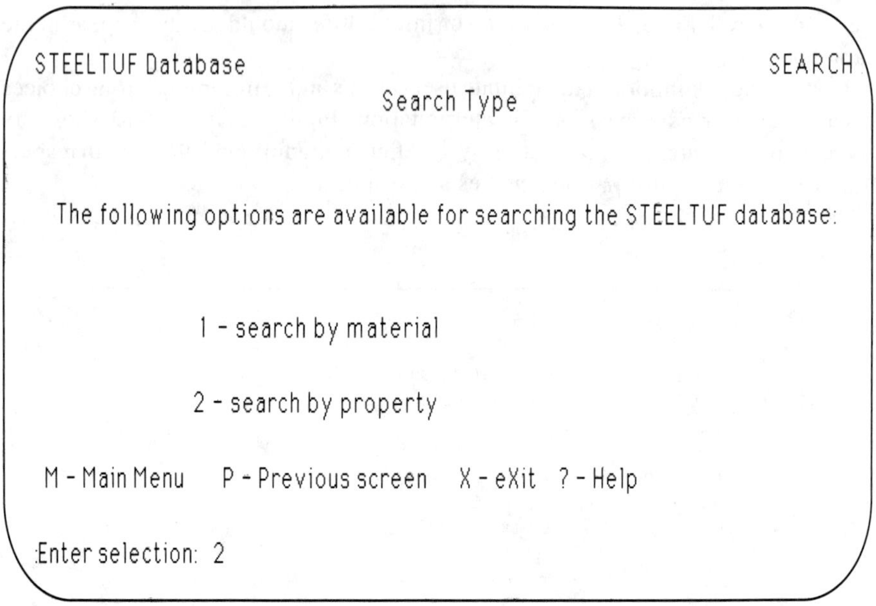

```
STEELTUF Database                                        SEARCH
                        Search Type


   The following options are available for searching the STEELTUF database:



              1 - search by material

              2 - search by property

 M - Main Menu     P - Previous screen     X - eXit   ? - Help

:Enter selection:  2
```

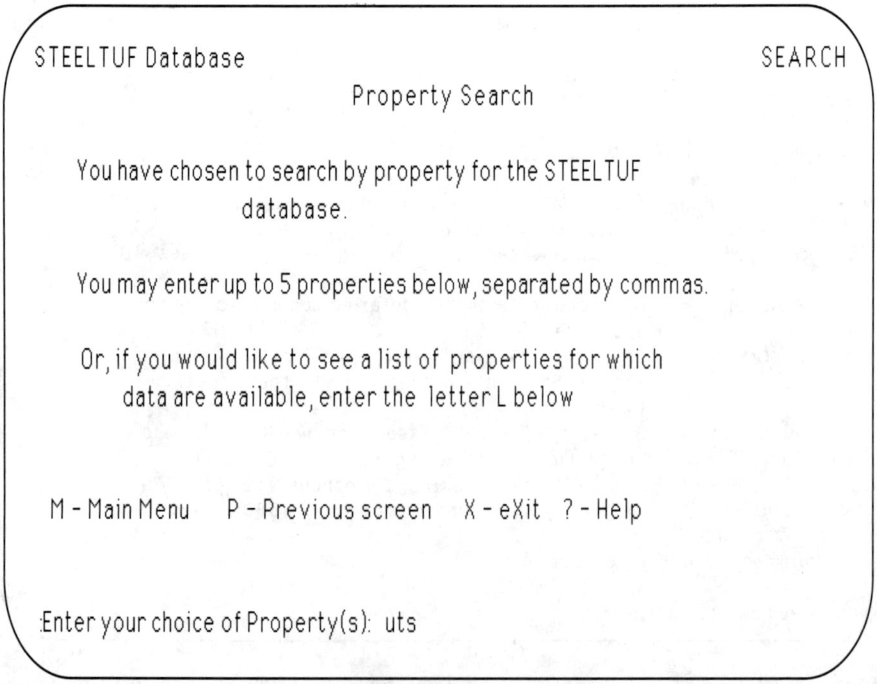

```
STEELTUF Database                                        SEARCH
                      Property Search

       You have chosen to search by property for the STEELTUF
                     database.

       You may enter up to 5 properties below, separated by commas.

       Or, if you would like to see a list of properties for which
              data are available, enter the  letter L below



    M - Main Menu     P - Previous screen     X - eXit   ? - Help


:Enter your choice of Property(s):  uts
```

Figure 5.11 These two screens continue the consistency as shown in Figure 5.10.

```
           STEELTUF Database                                SEARCH
6 Records

      Current search:

      Database:   STEELTUF
      Material(s): SA533A-2
      Properties:  UTS(350/800 MPa)

      Would you like to:
          1 - display results on the screen
          2 - change your request
             (This will keep your current database only)
          3 - download the results (not implemented  yet)

   M - Main Menu     P - Previous screen     X - eXit    ? - Help

:Enter selection:  1
```

```
Database: STEELTUF  Materials: SA533A-2  Properties: UTS
   Type:    Test  Data
   Result: 6 Records
Rec#  Material         Form              Test Variables        Dependent Variables
                                         (min/max)(units)      (min/max)(units)

 1    SA533A-2         Plate             TTEMP(172.0/366.0)    UTS(636/860)(MPa)
                                          (deg K)
 2    SA533A-2         GTA weld metal    TTEMP(172.0/366.0)    UTS(748/929)(MPa)
                                          (deg K)
 3    SA533A-2         GTA weld metal    TTEMP(172.0/366.0)    UTS(785/981)(MPa)
                                          (deg K)
 4    SA533A-2         GTA weld metal    TTEMP(172.0/366.0)     UTS(754/916)(MPa)
                                          (deg K)
 5    SA533A-2         Plate             TTEMP(294.1)(deg K)   UTS(655/717)(MPa)
 6    SA533A-2         Plate             TTEMP(294.1)(deg K)   UTS(696/758)(MPa)
No More Records in Display
Type N(Next) or P(Previous) screen or Q(Quit  Display)  or ?term = Help
for detailed display enter record number(s)  #,#...,#
Enter Choice: 1
```

Figure 5.12 Two final screens which remain consistent throughout display (see Figures 5.10 and 5.11).

2. Clarity

This is needed so the user experiences no confusion or uncertainty. Choices must clearly represent different functions. The system response must be distinguishable from the user response. Summaries of search strategies must be separated from search results. Keystroke options must be clear and unambiguous.

Columns of numbers and text must not run into each other. Header information must be different from repeating information. Field names should not be mistaken for data themselves. If printed reports are to be generated from screens, care must be taken to translate screen colours.

3. Crispness

The crispness of a display is an important design feature that helps centre the focus of the user on what is important and does not distract from the information. Database users often encounter complex screens full of characters with seven different colours, six different windows, or eight different shadings. Users are able to absorb only so much information at once and bombarding them with too much is just as bad as with too little. The amount of text should be limited per screen and information lined up. Colours should be chosen to guide or accentuate, not glare or distract. Window displays are confusing, not helpful, when too many are displayed.

4. Accuracy

Users also want and need accurate descriptions of expected actions. The system should do what it says it will do. Common words should not have esoteric meanings that confuse or mislead.

5. Logical layout

Finally, a logical layout must be employed. Important information should be located in important places, such as at the top, underlined or highlighted. White space or blank space on a screen is just as valuable as on the printed page for directing attention.

Almost the entire discussion of this philosophical approach to interfaces has been made without specific reference to technical data. These maxims represent the elements of good design and are equally applicable to interfaces for all types of databases. The sheer complexity of technical information often places additional demands on interface design, especially since so many elements of technical data are not easily represented, e.g., scientific notation and Greek letters.

5.4 TECHNIQUES OF BUILDING GOOD INTERFACES

Because we are discussing user interfaces, users must be involved from the beginning in their design. Even if the database builder is an expert user of the data, other users should be involved for review. Designers use many techniques in developing a user interface, and three of the most important will be discussed, i.e, paper copies, user meetings and prototyping.

1. Paper copies

Users will be most comfortable reacting to something concrete, and paper copies of proposed interfaces are useful because they can be marked up by users and designers to document proposed changes. For successful use, a complete set of the interface screens should be provided using word processors or screen designer software. Location, separators and windows should be included. Individual screens can easily be altered and the order changed as desired. The pages should be numbered to keep the sequence accurate.

2. User meetings

Paper copies can be passed out in advance for comments and collected at user meetings. At these meetings, transparencies of the screens can be reviewed collectively. The combination of individual comments with the consensus from the group greatly improves the efficiency of the process.

Design meetings have an important role to play in designing a database and developing a user interface (see also Chapter 9). Figure 5.13 gives a summary of rules on these meetings developed by Jon Pittman and Jan Veeder. These provide a good set of guidelines for successful meetings.

3. Prototyping

From these results, a prototype interface can be written for the first version of the database. This prototype should allow for rapid execution and realistic actions. The

Design Meeting Ethics

1. Do your homework

2. Criticize constructively

3. Try to achieve closure and maintain momentum

4. Best fit solution

5. Appropriate scale / appropriate context

6. Gracefully accept that 'It's handled'

7. Everyone is responsible for observing ethics

Figure 5.13 Design Meeting Ethics. (With thanks to Jon Pittman and Jan Veeder.)

review of the prototype interface must not be random but should be accompanied by written notes of any and all significant comments. After iteration, a good interface should emerge and alpha and beta testing can proceed. That is, users at the database building site (alpha) and then selected outside users (beta) react.

This process will vary considerably in actual application, but the essential elements can be identified: (i) initial written copy for review, (ii) good documentation of comments, (iii) a collective and synergistic review of the ideas and (iv) a prototype to test ideas.

When one or more of these steps are ignored or the designer assumes too much self-knowledge, poor interfaces result. The authors have seen many such cases and have participated in the design of some. By ignoring a step, the result is usually seen immediately upon release: the user interface is just not good enough.

While interface design is just one part of the entire database building process, it is one of the most important parts. Interface design must start as early as practicable and is often important in identifying key access paths that otherwise might remain hidden.

5.5   RULES FOR GOOD MENUS

Research on the design of good menus has been reported [1–3] where the emphasis is on empirical approach. Rules for good menus are therefore based on observation, e.g., on the users' view. Regardless of how well-thought-out or how similar to previous successful systems a set of menus might be, if the users do not accept them, they are not suitable. These rules apply equally well to menus presented in windows.

5.5.1 Menu choices

There are many ways to order the available choices on a menu:

   (i)   random order
   (ii) alphabetic order
   (iii) order in natural groups
   (iv) order of importance
   (v)   order of familiarity.

Of these choices, unless the menu items are totally unrelated, a random ordering is the least desirable. Alphabetic ordering is accommodating if the items are short and unique and the list not long. Contrast the choices in Figures 5.14 and 5.15. Alphabetic ordering in Figure 5.14 is accompanied by short symbols which add clarity. The long terms in Figure 5.15, though alphabetized, are more difficult to distinguish.

Several of the properties in Figure 5.15 are often called by other terms. For example, Young's modulus is also known as both the modulus of elasticity and the

For which alloying elements do you want to specify the concentration?

Carbon   C

Chromium   Cr

Molybdenum   Mo

Nickel   Ni

Phosphorus   P

Silicon   Si

Enter the appropriate chemical symbols, separated by a space or comma.

Figure 5.14 Alphabetized list in a menu. Symbols add to clarity.

Which Mechanical Properties do you want to search on?

bearing shear strength

fatigue crack propagation

plain strain fracture toughness coefficient

plastic fracture toughness coefficient

shear modulus

ultimate tensile strength

yield tensile strength

Young's modulus

ENTER CHOICE

Figure 5.15 An example where alphabetic listing is not a great aid.

elastic modulus. Thus, if the user is trained to use one of the alternative terms, the correct choice is not obvious. In Figure 5.16, the choices from Figure 5.15 are given in a natural grouping, affording the user better understanding of them, even though the terms may not be familiar to the user.

Young and Hull [4] have pointed out the types of problems that can be found in differentiating between menu items, including:

   (i) indistinct or overlapping items
  (ii) extraneous items
 (iii) conflicts or ambiguity
 (iv) jargon or unfamiliar words
  (v) generic terms.

5.5.2 Menu size

Menus can be set up in two ways, either with several screens (also called levels) with a few choices on each screen or with many choices on one screen. Empirical work

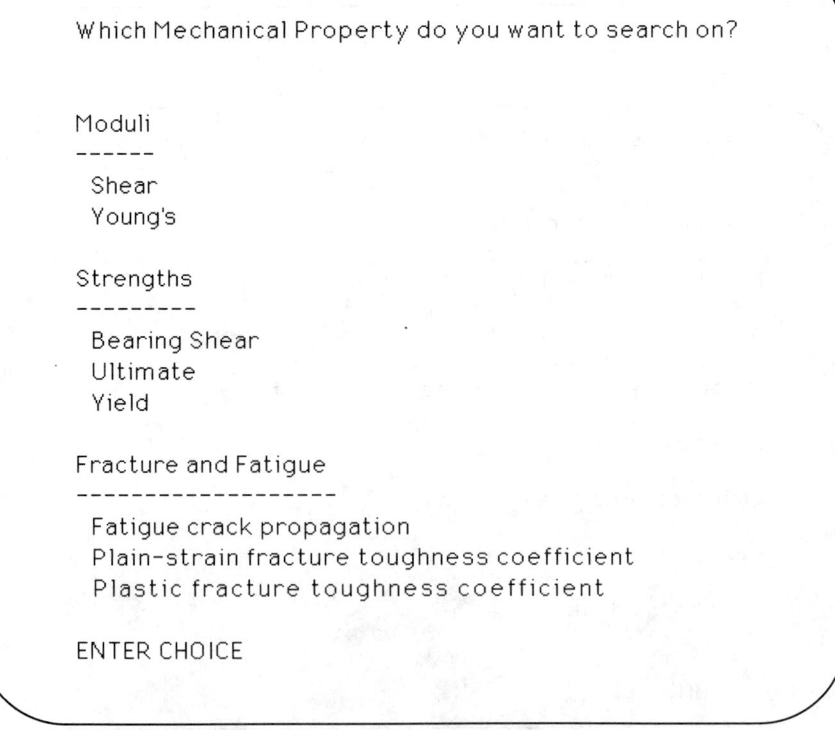

Figure 5.16 Natural grouping of choices in Figure 5.15.

has shown a preference to reduce the number of levels. Users seem more comfortable having more choices per screen and not having to select from many different menu levels. Seven appears to be a favoured number of items on one screen.

With the advent of windows, problems associated with a hierarchical menu structure are greatly reduced. Windows allow users to follow one chain of a hierarchy while maintaining contact with the main menu. Windows also allow for more precision and clarity on screens. Figure 5.17 shows the menu in Figure 5.16 using windows: the enhanced clarity can easily be seen. Window packages are available with many PC database management systems, and database builders can routinely include windows in their interfaces.

5.5.3 Selection in menus

There is no clear-cut preference for choosing a method of selection from a menu, and research has not shown any significant trends. Some important considerations have been noted. The use of numbers can be confusing and confining. In a long list, the middle objects tend to merge in the eye and wrong numbers can be chosen. Also, if a menu is changed and new items are added in the middle, the items following are renumbered and heavy users can easily make mistakes by relying on their past experience.

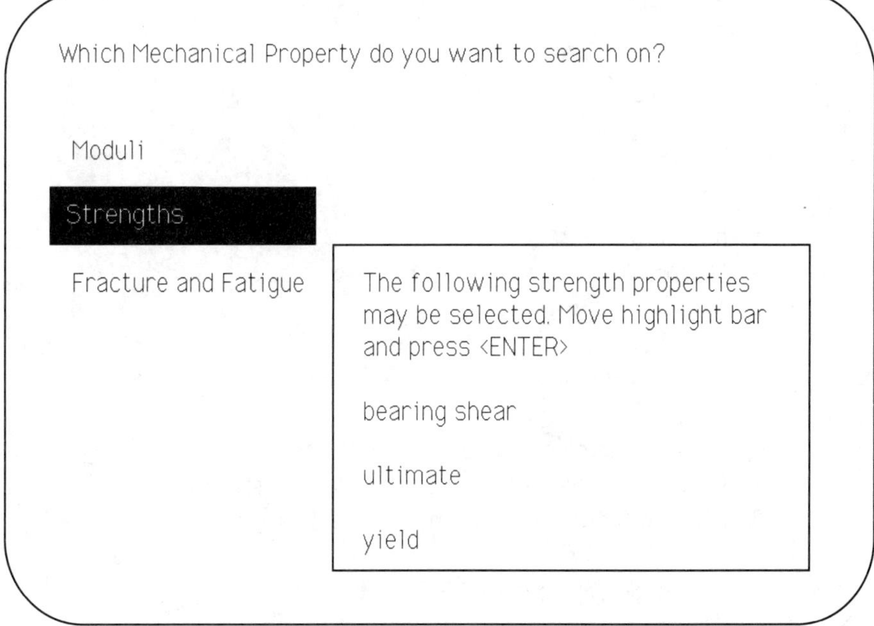

Figure 5.17 The menu in Figure 5.16 redone using windows. The contents of the window changes as different items are highlighted in the original menu.

Options include number and letter sets keyed to groupings. Figures 5.18 and 5.19 present examples so the reader can see the advantages and disadvantages. The use of mnemonics is appealing when there is clear distinction between different mnemonics (Figure 5.20). It is remarkable how often the use of mnemonics leads to collisions (that is, the same character string).

Newer computer terminals, PCs and mouses allow full screen addressing, thus allowing the user to highlight a choice or place a cursor against the choice. These techniques remove the ambiguity associated with numbers, letters or mnemonics and are superior (see Figure 5.20), particularly with the use of colour to emphasize the choice under consideration.

5.5.4 Navigation through menus

Once users become familiar with a database, especially if their usage is confined to a few predominant types of access or searches, they will want to pass through menus

Figure 5.18 Example of using numbers for menu selection.

```
For Which Mechanical Property do you want to search ?

Moduli

    a.  Shear
    b.  Young's

Strengths

    c.  Bearing Shear
    d.  Ultimate
    e.  Yield

Fracture and Fatigue

    f.  Fatigue crack propagation
    g.  Plain-strain fracture toughness coefficient
    h.  Plastic fracture toughness coeefficient

Enter letter of choice >
```

Figure 5.19 Example of using letters for menu selection.

```
For Which Mechanical Property do you want to search ?

Moduli

    shm   Shear
    ym    Young's

Strengths

    bs    Bearing Shear
    uts   Ultimate tensile
    yts   Yield tensile

Fracture and Fatigue

    fcp   Fatigue crack propagation
    psft  Plain-strain fracture toughness coefficient
    pftc  Plastic fracture toughness coefficient

Enter choice >
```

Figure 5.20 Example of mnemonics for menu selection.

quickly. Three techniques allow this to be done easily. The first is **typing-ahead**, that is, typing a response before the menu is fully displayed on the screen. A second method is **menu naming**, allowing the user to enter a later menu name and jump ahead. The last method is **menu macros or scripts** whereby the sequence of commands needed to reach a particular menu is stored and executed as a macro. Few technical databases have these features built into them now, but as user acceptance grows, these are expected to become more common.

Two important commands for any system are a **quit command** to exit the database and a **reversal command** to go back one step, usually after making an error. These should be the same for all screens of the database.

5.5.5 *Summary of menu topics*

In summary, menus are especially appealing because they reduce the amount of information a user must remember and provide rapid access without torturous typing of commands. Secondary actions can be easily accommodated and key information highlighted in a variety of ways. The elements of good menu design rest on a few basic principles. However, the user is the final arbitrator, and acceptance is the real test of quality.

5.6   COMMAND LANGUAGES

Command languages provide the primary alternative to menus for database users. Speed and efficiency are the major advantages of commands. Users are able to navigate through a system quickly, without displaying full-screen menus. As users become familiar with a system, this speed helps to reduce access time and to reinforce the users' control and understanding. A few years ago, when writing to a screen was slow, commands were preferable to save time and reduce waiting. The use of commands, however, has declined because terminal speed has greatly increased.

Command languages suffer from several severe problems, the most important of which is that commands are designed by nonusers and computer programmers. What may be beautiful to the designer by being cryptic, terse or brief, is usually obscure, confusing, or meaningless to a user. As with most knowledge, its structure of commands is often the key to widespread acceptance, especially by casual users. Commands usually contain three parts: the command itself, arguments, and options.

Commands are often single simple words that directly describe the action to be undertaken. A single letter or function key is also possible. **Arguments** in commands are usually files, data fields, etc that are to be acted on. They also can be abbreviated, but a prescribed order must be followed. Finally, commands often have **options** that provide variations in the actual operation, such as the number of copies of a particular file to be printed.

Some empirical rules, tempered by common sense, can be put forth on the structure and character of commands, options and arguments. Consistency in all commands is important, especially for order and structure. Abbreviations should be easy to remember and not too close to one another. Experience has shown that most users refuse to abbreviate commands for the first half-dozen sessions, after which they gradually do so. Generally users can remember only a small number of commands, so a quick display of possible commands and their meaning becomes necessary, which of course is a form of menu display.

Errors can be significantly reduced by the use of prompts, even if they are terse. For example:

```
Enter (Print, Copy, Delete, Save)
```

provides the user with enough information to take action.

For destructive actions such as deleting files and replacing data records, verify before executing. For especially destructive acts, double verification using different key strokes is suggested. Too many people have deleted important files just because they pressed the ENTER key twice in a row without thinking.

In spite of these problems, commands will continue to be used. As long as the designer maintains simplicity and keeps the user in mind, most of the negatives associated with commands will be avoided.

5.7 DISPLAY OF OUTPUT RESULTS

Discussions of user interfaces often ignore the display of final results. Output displays, however, are usually the only method of getting information out of the system until the data are printed and therefore are very important. For scientists and engineers, the conventions found in printed technical information sources, i.e., tables, graphs, diagrams and text, are what is ultimately required. Unfortunately, many of these conventions are not reproducible on present-day computer screens. Here we want to mention some aspects of displays related to interface design principles.

One of the most effective means of displaying results is to layer the output so the user has different options. Figures 5.21 and 5.22 show two screens from recently completed databases. The first screen presents a summary of search results and the second informs the user of further display options. The basic features should be mentioned. First, the search criteria used to generate these results are repeated. Then a brief summary of the database entries that satisfy those criteria is given. In many cases technical databases contain several levels of information or similar data from different sources that may not be displayed appropriately together. In these instances, considerable care must be given to what is summarized.

In the particular example shown in Figure 5.22, data from several databases have been found that meet the search criteria. However, one database contains unana-lyzed test data, the second contains data extracted directly from publications, and

```
      Matches from element and line search (total 125)
         element symbol  = Si
         line designation  = 2p


      Result           Line                              Phys.
      No.  Energy El  Designation   Compound             State

       1   101.6   Si    2p          (Me2SiO)5            VD
       2   102.6   Si    2p          (SiMe2)6O
       3   102.8   Si    2p          Al2Si2O7.2H2O
       4   102.8   Si    2p          Al2SiO5,kyanite
       5   103     Si    2p          Al2SiO5,mullite
       6   102.6   Si    2p          Al2SiO5,sillimanite
       7   102.63  Si    2p          Albite (NaAlSi3O8)
       8   102.9   Si    2p          Bentonite
       9   102.1   Si    2p          Et2SiCl2             VD
      10   101     Si    2p          Et3SiBr              VD
      11   101.4   Si    2p          Et3SiCl              VD
      12   101.8   Si    2p          Et3SiF               VD
      13   102.9   Si    2p          EtSiCl3              VD
      14   99.5    Si    2p          Fe3Si
      15   103.28  Si    2p          H Zeolon

      PageDown/PageUp        Main Menu: 0      Display: (Result No. 1-125)
```

Figure 5.21 A summary screen from the NIST X-Ray Photoelectron Spectroscopy database. Users can request additional information for any entry.

the third database contains definitive design information that has been subjected to rigorous scrutiny by a panel of experts. Therefore, the summary displays between Figures 5.21 and 5.22 take on a different character in spite of the similarity in the database searching techniques. Once a brief summary is given, the user may wish to see complete information for a given data set. Other aspects of display have been discussed in Section 4.4.

5.8   COLORS

Colors are a computer display's answer to a high-quality printing press, allowing information to be highlighted or downplayed just as type fonts, point size and white space are used on the printed page. The first attempts at color were often excessive, with full rainbows flashed across the screen much faster than any user could appreciate. The subtle use of color, resulting in much more appealing technical databases, has been very effective. Persons interested in technical databases should

```
  Properties: KIC
    Result: 160 Records
  Database: ASMH Type: Literature Citation
  Rec# Material      Form           Test Variables          Dependent Variables
                                    (min/max)(units)        (min/max)(units)


    1   4340          Plate                                 KIC(60/89)
                                                            (ksi.m**0.5)
    2   17-4PH        Bar                                   KIC(54/100)
                                                            (ksi.m**0.5)


  Database: STEELTUF Type: Test Data
  Rec# Material      Form           Test Variables          Dependent Variables
                                    (min/max)(units)        (min/max)(units)


    3   A387-2 CL. 22  Plate        TTEMP(123.0/183.0)
                                    (deg K)
    4   A387-2 CL. 22  Plate        TTEMP(113.0/223.0)      KIC(32/192)
                                    (deg K)                 (mpa.m**0.5)
    5   A36           Plate         TTEMP(171.8/227.4)      KIC(45/132)
                                    (deg K)                 (mpa.m**0.5)


  Records Continue on Next Page
  Type N(Next) or P(Previous) screen or Q(Quit  Display)  or ?term = Help
  for detailed display enter record number(s)  #,# ...,#

  Enter Choice.
```

Figure 5.22 The summary screen from the MPD Network. Compare to Figure 5.21. Different information is shown including independent variables and database source.

be aware that a standards activity relating to the use of colors in interfaces has started under the auspices of the International Standards Organization. Much of this effort is aimed at using color to promote awareness in important situations and ensuring that different countries do not have different conventions.

5.9 SUMMARY

The user interface is a major feature of a technical database because it represents what the user actually sees. Progress in the last decade has identified important features that all good interfaces have in common. Rather than restricting creativity, this awareness has allowed interface designers greater freedom to develop novel and interesting interfaces which also function well. Good interfaces enhance databases; bad interfaces inhibit their use.

REFERENCES

[1]   G. Kearsley and R. Halley, *Designing Interactive Software*, Park Row Press, La Jolla, CA, 1986.

[2]   *The Design of Interactive Computer Displays - a guide to the select literature*, Eds. K. McGee and C. Matthews, The Report Store, Lawrence, KA, 1985.

[3]   B. Shneiderman, *Designing the User Interface*, Addison-Wesley Publishing Co., Reading, MA, 1987.

[4]   R. M. Young and A. Hull, Cognitive Aspects of the Selection of View data Options by Casual Users, in *Pathways to the Information Society, Proceedings of the Sixth International Conference on Computer Communication*, London, 1982.

[5]   J. Johnson et al., The Xerox Star: A Retrospective, *Computer*, 22, 11–29, 1989.

Chapter 6

The Dissemination of Technical Databases

6.1 DISSEMINATION OVERVIEW

Databases are intended to be used, and the dissemination of their data is an important part of the entire building process. Dissemination is not, however, announcing a database to the world, sitting back, and waiting for the world to beat a path to your doorstep. Dissemination alternatives must be actively considered at the beginning of the building process because they require different actions. The primary options are:

   (i)   personal computer packages
   (ii) packages for midsize and large computers
   (iii) online systems
   (iv) packages in software or instruments

Each has advantages and disadvantages, and the choices depend on the user community, application and the kinds of computers the users will have.

6.2 PERSONAL COMPUTER PACKAGES

The widespread availability of personal computers has opened a tremendous market for technical databases. PCs now allow databases to reach end users, totally under their control. Their self-contained nature makes them attractive; generally the user simply has to put a diskette into a PC, type a few commands, and the database is loaded and ready to use. Occasionally additional software such as graphics is required. PC's also allow for appealing user interfaces (Chapter 5).

PC databases are distributed on a variety of diskettes, floppy (5 $\frac{1}{2}$" or 3 $\frac{1}{4}$"), hard (e.g., for Bernoulli boxes) or optical (CD-ROM). There are also different types of PCs and different operating systems. Many combinations confront the database vendor, and usually only a few configurations and diskette types are supported. The diskettes, distributed in a sturdy container for protection and storage, include a user's manual.

If the database uses a commercial database management system, suitable licensing agreements must be made. Most DBMS vendors do allow third-party distribution of a run-time version of their product for a small fee. A key consideration is whether the user will be allowed to make changes in the database, such as adding new fields or additional data. If this is the case, some vendors feel that the users are no longer using just a run-time version, but instead are doing their own database management, and the vendors charge the full licensing fee. Negotiations between the DBMS vendors and database builders are complicated and highly individualistic. No special guidance can be given except to say that, as in any business arrangement, it is important to settle issues before work has progressed too far. Of course, home-built databases can be disseminated without such licensing problems.

6.3 PACKAGES FOR MIDSIZE AND LARGE COMPUTERS

Often databases are intended for use on mainframe computers, allowing access by numerous users. In this case, magnetic tapes have been the medium of choice to date since magnetic tapes have been well-standardized and hardware and operating system problems are rare. With the widespread availability of local area networks, uploading databases from a PC is possible, and the database can be sent on floppy disks.

The primary differences between packages for the PC and those for mainframes are the way databases are used and maintained because of the absence of direct control by users on a mainframe. Starting and stopping the software and the handling of errors, as well as file management, take on new dimensions. The licensing of DBMS software packages can also be more complicated on a mainframe. Nonetheless, mainframe packages are appealing to technical users because of the large number of VAX-type machines in use.

6.4 ONLINE SYSTEMS

A database can be made widely available by an online system (Figure 6.1), often through a third-party vendor. Database builders must work closely with the vendors to achieve compatibility. A database may need to be modified substantially to make it suitable for the vendor's system, a process which takes time, heavy involvement of the builder and a high cost. Rarely is a database so attractive financially that a vendor does this work free. The usual pattern is for the database builder to waive royalties for a period of time, rather than actually transfer funds. The transformation work can easily take six months for a midsized database, and database builders should be aware of online vendors who claim this can be done trivially. The authors have not seen this happen in practice.

Four important factors must be considered in adding a database to an online system:

(i) correct interpretation of the data; each data field must be understood and handled accurately,

(ii) determination of the equivalence of data and data fields with other databases on the system,

(iii) transformation of the physical data structure to a new DBMS,

(iv) display of the data on the online system.

This work requires considerable interaction, especially at first. If possible, the database should be prepared in three stages: (i) on a 1–2% level that is representative, (ii) a 10% level that samples the entire database, and finally (iii) the entire database. The smallest file should be forced through the entire process quickly. In this way, the magnitude of the loading process can be understood and major issues identified early in the effort. The 10% file can then be used to test proposed changes.

There are advantages to making a technical database available via both an online system and as a standalone system on a PC. However, some sophisticated database capabilities may simply be beyond the reach of a small database. For example, chemical substructure searching capability might not be feasible for a database that

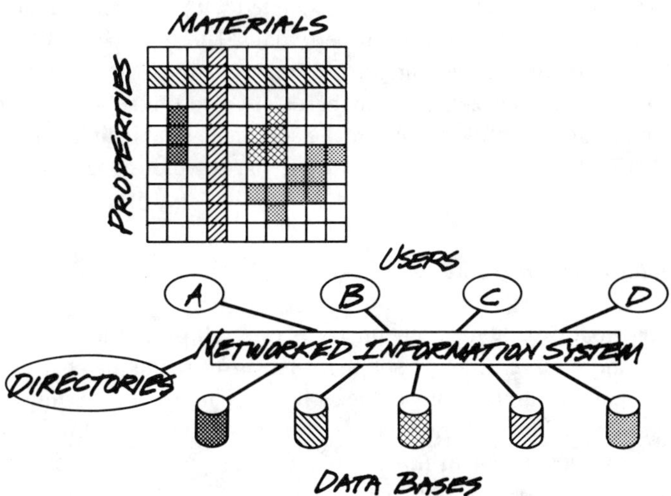

Figure 6.1 A database network is being developed for materials property data. Each database contains a different selection of materials and properties as represented by the shading in the matrix.

contains only 10 000 compounds and 10 to 20 data fields. An online system shares the costs, both developmental and operational, over all databases on their service.

A problem with PC databases is that databases are often not adequate to solve a typical technical problem because they do not contain all the required information and it is very difficult to combine data from different PC databases. However, online systems can provide access to information from several data sources together.

6.5 DATABASES INTEGRATED INTO OTHER PROGRAMS AND INSTRUMENTS

The final mode of dissemination for a technical database is to make it part of some package. The most prominent examples of this are inclusion of databases in larger software packages or as part of an instrument. An example of each will be given to illustrate the possibilities.

6.5.1 Databases as part of a software package

Chemical engineering has been revolutionized by the advent of computer modelling of chemical processes. Rather than resorting to expensive experimental scale-ups of promising new manufacturing processes, chemical engineers can now call upon sophisticated computer models for the mass and heat flow in a process. Every major chemical company and many universities have built such software.

Most of the models have a database associated with them that includes thermo-chemical as well as other physical property data. These property databases are in some cases disseminated separately, but are used most heavily as part of the software package. Other technical databases are also linked to the software. The group maintaining the software must make sure that suitable links to the databases are included so that data can flow from one part of the modelling procedure to another.

6.5.2 Databases for instruments

The primary goal of analytical chemistry is to establish the identity of unknown substances, and a large array of sophisticated instruments has been designed to make this science accurate. Many of the techniques involve examining components of the unknowns, either by fragmenting the entire species and using mass spec-trometry or by viewing them directly via other types of spectroscopy. Various measurements are made, and with pattern recognition techniques the identity of the components are determined, which when combined can establish the identity of the original unknown substance. Before the use of computers in instruments, this identification was a lengthy process done by chemists using published data collections.

Nowadays, the pattern-matching techniques are computerized and the results matched against large databases of previous measurements. Mass spectrometry is one analytical method that has been automated, and several different databases and identification software packages are now in use. Because mass spectrometry is a primary tool used to identify pollutants in the environment and is a technique which has been adopted by virtually every analytical laboratory in the world, the market for mass spectral databases is extremely large. Instrument manufacturers have developed the needed software, adapted the databases as necessary and paid royalties for each copy distributed. Many other databases are similarly treated in areas such as infrared spectroscopy, powder diffraction analysis and electron diffraction. The potential to reach large audiences is great.

6.6   SUMMARY OF DISSEMINATION TYPES

In concluding this discussion, we reiterate that for a database to reach the widest possible audience, a variety of dissemination methods can be used. Probably a combination of several is appropriate. Although PC databases are presently viewed as glamorous, other types of dissemination cannot be ignored.

6.7   PRACTICAL CONSIDERATIONS IN DISSEMINATING DATABASES

Even though databases contain some information equivalent to that in books, many relevant issues are still unresolved. These include copyright, liability, leasing agreements, downloading, copy control, pricing, updating, user support and documentation. Each of these issues will be discussed briefly. Kaufman [1] has discussed related issues from the viewpoint of materials databases.

6.7.1 Copyright

The basic purpose of the copyright law is to ensure protection of an author's rights with regard to a created work. A database is considered a created work and is thus covered by copyright laws. Copyright protection of a covered work begins at the time of creation, but to facilitate enforcement of the laws, copyright registration is strongly encouraged and registration procedures are well-established. Unlike books, works of art, music and other protected works, computer databases are new to the copyright world, and this situation has two consequences. First, there is a lack of case-law history providing interpretation of appropriate laws. Second, most national copyright offices are not familiar with databases. As a practical matter, the second of these is changing rapidly. The first issue is a problem because the legal advice given today about database copyright concerns is based primarily on conjecture, rather than experience. When lawyers have to conjecture, they are usually cautious.

The appropriate time to register the copyright of a database is when the dissemination copy is finalized. When changes and updates are made, a new registration should be made. For example, databases that are updated yearly have their copyrights registered each year.

A copyright notice must appear on each copy of the disseminated database as well as on the documentation. This is usually done on the opening screen, so users must see it each time the database is used. For online systems the copyright notice for individual databases should appear upon access. A sample is shown in Figure 6.2.

NIST Standard Reference Database 20

NIST X-RAY PHOTOELECTRON SPECTROSCOPY DATABASE
Version 1.0

Data compiled and evaluated by
C.D. Wagner
Surfex Company

Program written by
D.M. Bickham
Standard Reference Data NIST

Distributed by
Standard Reference Data
National Institute of Standards and Technology
Gaithersburg, MD 20899

Copyright 1989 by
U.S. Secretary of Commerce
on behalf of the United States Government

** Press any key to begin **

Figure 6.2 The copyright notice from a database distributed by the National Institute of Standards and Technology. For nongovernment groups, the notice would be shorter. This notice appears every time the database is started.

6.7.2 Liability

Questions about liability for technical databases, for both database builders and vendors, should be referred to lawyers for advice. Again there is minimal case law associated with the use or abuse of technical databases. Initial opinions, formal and informal, have emphasized that good scientific and engineering practices must be followed whether the data source is computerized or published. One special concern is that cautionary information often displayed on printed pages (footnotes, parenthetical expressions, etc) must be equally prominent when the same data are shown on a screen.

6.7.3 Leasing and sale agreements

When you buy a book, you pay your money, and receive the book. There is no written sales agreement, no contract to be signed, no prohibition on lending the book to a friend, etc. This is because many years of experience have been accumulated in copyright laws for books and in the protection of the rights of authors.

For databases, lease and sale agreements have become the norm and are rapidly approaching the size and complexity of the average home mortgage contract. Two reasons primarily account for these voluminous agreements.

First, for large organizations that employ many scientists or engineers in various locations, the possibility of purchasing one copy of a database and making it available to everyone is real. A trend exists in the software industry to provide site-wide and organization-wide agreements, and technical database dissemination is following the same practice. These agreements usually establish a per copy price as a percentage of the single copy price. In addition, after a certain number of copies have been acquired, unlimited copies can be made. Another related problem concerns the issue of what constitutes one copy, especially with the proliferation of local area networks and distributed database systems.

The second concern is that the database will fall into the hands of an outside user who will consider it as freeware. While this is unlikely in a corporate environment, the chances are very real in student environments. Some technical database builders are interested in the student market, either with condensed versions of databases or with diskettes associated with textbooks.

6.7.4 Downloading from technical databases

Downloading (Section 4.3.1) from online systems is going to happen, so there is no point in fighting it. The price structure of the online service must take such downloading into account. The danger is, if an entire database is downloaded, the service is not needed again. Large bibliographic online services have already built techniques to monitor such abuses, and these can be added to numeric systems as well.

6.7.5 *Pricing of technical databases*

Prices for technical databases, as also with leasing arrangements, differ considerably from those for books by being much higher. Presently, this is because competition is lacking for most technical databases. Databases certainly can provide more capability than books, and producing a technical database is more costly. Obviously prices in the software market have fallen dramatically in recent years, especially as new packages with more capabilities become available. There is every reason to believe that the same pattern will be followed by technical databases. In addition, once the initial database package investment has been recovered, additional data sets using the same software will be priced lower. This is already the case with several materials databases.

6.7.6 *Updating*

Several years' experience now lead us to believe that the rate-determining (slowest) step in the updating of technical databases is the technical work itself. Continuous updating of these databases is not normal nor is it needed. In most cases, the data themselves remain rather static, and changes and additions are only incremental. Many technical databases are now updated on an annual basis, leading to a well-defined work effort as well as easily marketed update products.

When technical databases have been incorporated into other software or changed in some fashion by individual users, incorporation of updates is a bigger problem.

6.7.7 *User support*

One successful database manager told us that after disseminating the first 400 copies of his database, he received 400 phone calls. Some calls were just to say that the user successfully installed the database; others were more substantive. The primary reasons for calls from users are given in Table 6.1. Perhaps as database builders get better at making truly easy-to-use products, such communication from users will decline.

Regardless of the reasons, database suppliers must expect that users will want to be able to talk to a human being, especially at the start. A telephone number where

Table 6.1 Reasons why users call database vendors.

Poor instructions on how to get started
Unexpected failures or responses
Unclear error messages
Lack of expected data

a capable person can be reached is crucial to satisfy this need. For more online systems, user support normally includes training and marketing. The Chemical Information System, while it was still a government system, found that about 20 percent of its costs were allocated to such activities. Private vendors cite similar figures.

Very few technical databases have a users group. These groups can be beneficial, not only in spotting problems, but also in suggesting new directions.

6.7.8 Documentation

Technical databases are getting better, but the record for documentation is still mediocre. For technical databases, five distinct levels of documentation are needed, normally already provided. Unfortunately, they are not done well enough. The documentation required is shown in Table 6.2.

The **installation instructions** must be foolproof and cover all anticipated situations. A phone number should be included for the unexpected problem. The absolute best test for the quality of these instructions is to give them to nonusers and see how well they do.

User manuals are unfortunately growing in size rather than in clarity. What is needed are concise manuals with easy-to-find features. The manuals should include examples of successful use, so that the user can gain confidence. Good subject locators are equally important. These manuals should be written in everyday terms, not in jargon. Again, the best test is to give the manual to a new user for review.

Action summaries are usually small cards that provide a synopsis of commands or actions. The usual alphabetic ordering is poor, especially if the name of a command is not known. Grouping actions by categories and having duplicate entries are good ideas.

When using a database, users expect **online help** to indicate possible actions, to explain those actions, to define data fields, etc. Online help should not take pages when a sentence or two will do.

Finally, an **archival technical publication** explaining the technical basis for the data collection is needed for each database. These articles should be comprehensive

Table 6.2 Documentation required for a technical database.

Installation instructions
User manual
Summary of common actions and commands
Online help
Technical publications

but not overly technical. References to more detailed work can be given. These articles should be noticeably cited in the user manual, among other places.

6.8   ACCESS TO TECHNICAL DATA SOURCES

Up to this point, this chapter has discussed technical considerations and business details important for successful databases. We now focus on aspects of how users actually acquire databases, including how they learn of their existence and how they acquire them. This discussion will cover several problems with the dissemination of a technical database. After the creation of a technical data source, scientists and engineers access the data in a four-step process, as shown in Table 6.3.

Figure 6.3 compares the steps for published hard copy (books, etc.) and published computer databases. Database access involves everything under hard-copy publications (left side of Figure 6.3) plus additional problems (right side of Figure 6.3). As the third and fourth steps are detailed in other parts of this book, we will concentrate on the first two steps here.

Table 6.3 Stages of accessing a technical data source.

1. Awareness

2. Acquisition

3. Retrieval

4. Extraction and use

6.9   AWARENESS OF TECHNICAL DATABASES

The problems of accessing technical databases begin with awareness. To even consider using a data source, the user must know that it exists and what is in it. This has been one of the biggest barriers to using published hard-copy data sources. Publishers do extensive marketing, libraries maintain catalogues, and many groups develop comprehensive computerized services to help technical workers identify relevant publications. Even though these have been remarkably effective, there are still many instances where important hard-copy data sources are relatively un-known.

Marketing and publicity are necessary for databases, and the vendors are using techniques similar to those of book publishers. Because databases often contain significantly larger amounts of data than books, a detailed index to their contents is needed and can be created with a precision that bibliographic abstracting services

	Publications	Databases
Awareness	Catalogs Word of Mouth Libraries Table of Contents Indexes	Directories Indexes Online Catalogues Data Thesaurus
Acquisition	Purchase Loan Copy	Purchase Lease Online Access
Retrieval	Table Reading Text Reading Graph Reading Written Language	DBMS User Interfaces Nomenclature Query Language
Use	Writing Reports Making Tables Drawing Graphs Copying	Report Generation Downloading Graphics Modelling Software

Figure 6.3 Steps in accessing technical data sources. Each step consists of several activities; those for databases also involve all those listed under the Publications column for the same step.

lack. There was one attempt to create such an index which unfortunately did not seem to work. Online services could offer a comprehensive central catalogue of these indexes, thus avoiding the piecemeal publicity now used. Until such marketing tools are implemented, dissemination of PC databases will encounter the same problem that books have now. Many users do not know of their existence.

6.10   ACQUISITION OF TECHNICAL DATABASES

Any PC database builder can claim that its product is inexpensive and not difficult to purchase. However, scientists and engineers cannot buy an unlimited number of PC databases, any more than they can buy an unlimited number of books. As pointed out, users often need more than one database in their technical work. Already several basic facts about technical databases have become obvious, as summarized in Table 6.4.

Methods will be needed to allow the user to acquire and use many databases easily and concurrently. Possibilities include online systems with or without database integration or PC database packages built to standards so that many different data sets can be used; alternatively CD-ROMs containing many databases may be accessed. What the future trends will be is certainly unclear, but some means of integration will be necessary.

Table 6.4 Basic facts about technical databases.

The same people will build databases as those who write books
Most groups will not want to provide access via their computers
Most technical people will soon have a computer at their desks
Most technical people use a multiplicity of data sources
Most people do not want to learn to use different types of computers

6.11 SUMMARY

Many options are available to technical database builders in terms of disseminating their products. The optimal choice depends on the scope of the database, the type of user likely to need it and the size of the potential market. All the options require attention to practical matters including copyright, documentation and support. These options cannot make a bad database good, but ignoring them can make a good database bad.

REFERENCE

[1]   *Guide to Material Property Database Management*, Ed. J. G. Kaufman, CODATA Bulletin, No. 69, CODATA, Paris, France, November 1988.

Chapter 7

Data Structures

In the previous chapters we have discussed the nature of technical data, their use and application, as seen from the point of view of a scientist or engineer who might use the data. In this and the subsequent two chapters we turn to the computer database systems used to process the data, the structures used to store the data, the structure or architecture of the whole database systems and the various models used in most database systems. In this chapter we consider the structure of the data.

In any database system, the data must be structured to enable us to retrieve the information we need from the database as quickly and as efficiently as possible. To understand the possible data structures we begin by describing the storage media, disks and tapes on which the data structures are based.

Many books on database systems include at least one chapter on data structures. A useful overview is given by Elmasri and Novathe [1] and a comprehensive review has been published by Semet [2].

7.1   STORAGE MEDIA

Although cards, paper tape and magnetic cards have been used to store some databases in the past, and semiconductor memory will probably be used in the future, databases today are held on magnetic disks, optical disks and magnetic tapes. We discuss these separately below.

7.1.1 Magnetic disks

There are two types of magnetic disks: hard disks, and floppy or flexible disks. The floppy disks, used so frequently with personal computers, can support only small limited database systems at present. Nevertheless, the capacity of these disks is growing and it is possible to hold personal databases, or databases for small projects or experiments, on one or two floppies. However, this cannot be recommended because of their relatively poor performance compared to hard disks, typically by a factor of 10, whilst hard disks are only slightly more expensive.

In this chapter we therefore concentrate on the hard disk. Normally this is a fixed disk, built into a sealed unit in the disk drive, protecting it from the atmosphere, and it is normally not removable. For example, on a PC we can buy such a small hard disk for a low cost, with a capacity of 20 megabytes upwards, and the prices are falling while capacities are rising each year. Hard disks vary in size from $2^{1}/_{2}$", $3^{1}/_{2}$" or $5^{1}/_{4}$" diameter on PCs to 14" on mainframes. The great bulk of database systems, even on PCs, will be based on hard disks; floppy disks will still be used on these systems, but for holding back-up copies of programs and for data interchange.

The device for reading from or writing to a disk is called a disk drive. Data are stored magnetically in circular tracks on the magnetic oxide coating of each disk surface using a read/write head. Normally the same head is used for reading as well as writing. The head can move to align itself over any one of several tracks on the disk, making it possible for the same read/write head to access any data on the disk (see Figure 7.1). The data are stored in blocks (sometimes called pages), each typically 512 or 1024 bytes in length, a size suitable to be transferred between the disk and a buffer of the same size in the central store of the computer. All data are transferred to and from the disk in these blocks or in multiples of blocks, sometimes also called 'buckets' of data. A horizontal check digit, usually a sum check, is used to ensure that the data are correctly read from the disk or written to the disk.

To read (or write) a block on the disk the read/write head must first move to the correct track, and the 'seektime' required to make this head move is typically between 10 and 100 ms, depending on how close the head is to the track already and the type of disk drive. An average figure for a hard disk might be 30 ms. The head then has to wait while the disk rotates, until the correct block appears beneath the head. This may take another 10 ms. For most fixed disks you can count on a total

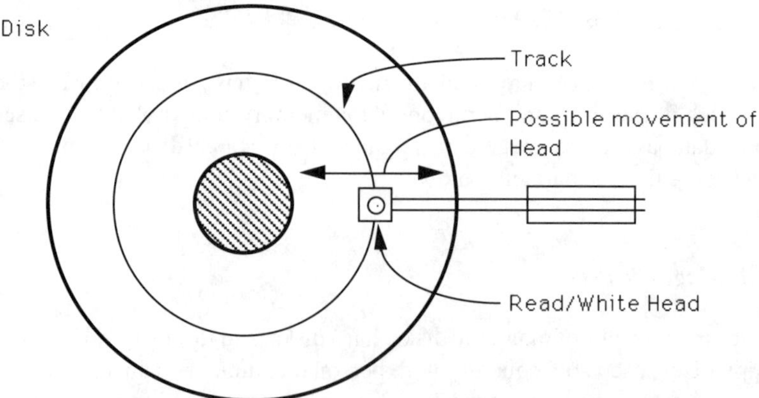

Figure 7.1 Outline of track and read/write head on a disk drive. The head can read or write data in blocks from the track directly below it.

average delay, or access time, of about 40 ms, except on the most expensive disks where this is about 10 ms. In human terms 40 ms does not seem to be a long time; it is only 1/25 s. But for a computer processor, which can execute a million instructions in a second, it is a large factor greater than the time it takes to process the data on the block; so if several blocks are being processed simultaneously the computer will spend a large fraction of its time waiting for the next disk access to be completed. It is therefore no surprise that most database systems are disk bound; that is, their performance is almost totally determined by the access speed of their disks; when busy there are often queues of processes waiting for access to each disk.

It is interesting to note that in the last 20 years processor power has improved by at least a factor of 1000; disk capacity has improved by a similar factor, but the access times of disks today is only slightly better than the access times of the first disks used with computers in the 1960s. Databases, which rely heavily on moving large amounts of data to and from these disks, are affected particularly adversely by this restriction. It is therefore very important that the database software be designed to minimize the number of accesses to the disk in the performance of each task. The data structures we discuss later are chosen with this end in view.

Although the time needed by the disk to read or write a block is about 40 ms, this unfortunately does not mean that your user program can execute 25 accesses in each second. In a multi-user environment it is likely that other users are also queuing to access the same disk and after each time you read a block (unless you have a very high priority) you will go to the back of the queue and have to wait your turn again to access your next block. But even if you are the only user, you will unfortunately still be unable to get 25 disk accesses per second. The reason is that the operating system, which bestows many benefits such as making your computer much easier to use, exacts a heavy price in the form of a poorer performance. We have found in practice that 3–5 disk accesses per second is often as many as you can expect on a single user machine, and one disk access per second is not uncommon on a busy machine.

In summary, it is disk accessing, above all, which slows down system performance, particularly database system performance. Structures to reduce disk accesses are therefore essential to a good database.

Disk capacity is no longer a problem. Advances in technology such as thin film heads, plated media and vertical magnetic recording have pushed up recording density and reduced costs, particularly on small computers; so you can now purchase a drive storing 500 megabytes for a few thousand dollars. This would be sufficient for all but the largest technical database not storing images or continuous signals. (These need to be stored on an optical disk.)

7.1.2 *Optical disks*

The power of optical disks [3] to store data arises from the properties of light and the development of the laser. Because a laser has one precise frequency it can be used to focus light to a very small point on the surface of a flat disk and heat a tiny circle only 1 micrometer in diameter. This can be used to change the state of the surface within the circle (the technology used for erasable disks) or it can simply burn a hole in the surface (the technology used for WORM disks and compact disks). Data are stored by recording a binary 1 as a change of state of the circle and binary 0 by no change in state. The circle storing the bit is much smaller than the domain needed to store a binary bit magnetically. It is thus possible using this technology to pack binary bits a few micrometers apart on a disk surface and therefore to store about 1 gigabyte of data on a $5^1/4''$ disk. This is a factor of ten better than is likely to be possible with magnetic disks.

There are three main types of optical disk commercially available: the WORM, the CD-ROM and the erasable M-O disk.

1. WORM

The term WORM is an acronym for **Write Once Read Many** times. Because the device depends on a technology which burns a small circular hole in the surface of the disk to represent a binary 1, it is not possible to reuse the disk a second time for new or modified data. So data are written only once to each area of disk, but they can be read many times. This limits the usefulness of the disk. However, it can be used for static data such as images or for holding whole copies of a static database. Normally the database is first built on a magnetic disk, edited and corrected, until it is in its final form and then copied in its entirety to the optical WORM. It cannot be edited again unless a complete new copy of the database is written to a different part of the WORM. Since the disk may hold about 1 gigabyte of data, this is often possible.

Once a database is copied to a WORM it is accessed in the same way as a database on a magnetic disk. Normally you can use it immediately after you copy it from the magnetic medium. However, it is about a factor of 4 slower than the magnetic version. The reason is that the read/write head on an optical disk is still heavier than the head on a magnetic disk and the increased inertia slows down the access time. So the performance is poor, particularly when writing data to an optical disk, since this needs more power in the laser.

These drawbacks, particularly the inability to write more than once which prevents you from editing data on the disk, mean that a WORM will not replace your magnetic disk drive. It has three main areas of application.

(i) **Images and drawings**. The storage of digital images, drawings or other continuous signals such as speech, requires a high amount of mass storage and would be too expensive on a magnetic disk. For storing an image which does not need to be changed, a WORM is ideal. For storing an engineering drawing a WORM is also suitable, as original and amended drawings can both be stored and often an engineer will need both.

Unfortunately there is still a great shortage of software to make these facilities available in database systems on the market. However, by numbering the images or drawings and by storing the numbers in a conventional database system, images can be retrieved quickly using a table mapping the numbers onto the addresses of the images. Although not ideal, this can be effective.

(ii) **Archiving infrequently accessed databases**. When a static database is only accessed occasionally it is too expensive to store it permanently on a hard disk, space on which is always in demand on every computer system. The database can then be copied, perhaps with other similar databases, to an optical disk. This is the second use of a WORM. Although subsequent access to the database will be slightly slower than access when it is on the magnetic disk, this will be acceptable if it is infrequently accessed.

If the database has to be changed it must first be copied to the magnetic medium, modified and then copied back. This long process would not be acceptable if frequent changes to the database were needed. Thus, this use is restricted to databases which only have to be updated occasionally. Technical databases often fall into this category.

(iii) **Dissemination of information**. The third use of a WORM is for limited dissemination of the information on a database. The database can be copied onto a WORM and the whole database sent to another installation or user and there accessed directly in its optical form or by first copying the whole database to magnetic disk and then accessing the magnetic disk version. This is suitable for limited dissemination of the database, say to tens of user installations. If the database has to be sent to hundreds of installations then a CD-ROM is more suitable. (See the next section.)

2. CD-ROM

The term **CD-ROM** is an acronym for **Compact Disk-Read Only Memory**. This uses the $5^1/4''$ compact disk widely available for high-fidelity digital sound. It is based on the same technology as the WORM, tiny holes in the disk surface representing binary 1; however, it is designed for the publication of huge amounts of information on a disk. A database is first built on a large magnetic disk. This is copied to a master copy of a disk (similar to a WORM) which is used subsequently to press large numbers of CD-ROMs holding the database, employing the same technology for pressing compact disks. Since this process is expensive (about \$20 000) it is only economical for the publication of 100 – 500 copies of a database system, not a common requirement for a technical database system. Nevertheless there are a large number of database systems available on CD-ROMs, including several crystallographic and electron diffraction databases [4].

3. Erasable M-O disk

The term M-O disk stands for **Magneto-Optical** disk and is one of the forms of erasable optical disk (the others, not so fully developed, are phase change and dye polymer disks). Erasable means that data can be erased from the disk or from any part of the disk, and the disk used again. So this allows the data to be edited and changed on the disk.

The M-O disk is based on a simple physical principle that magnetic particles easily change their magnetic orientation if they are heated beyond a certain temperature (the Curie temperature). The particles in a thin magnetic film are all magnetized in a certain direction. A weak applied magnetic field changes the magnetic orientation within the tiny spots heated by the laser. This change can be detected by a change in the polarization of laser light later reflected off the surface.

This process of writing data can obviously be reversed, producing an erasable optical disk. Although this gets over the main difficulty with a WORM, the other difficulty of poorer response time remains. So an M-O Disk is unlikely to replace your magnetic disk soon unless you need very large capacity.

However, it can be used for any purpose in place of a WORM and it is particularly useful for taking regular security copies of a large database system.

7.1.3 Magnetic tapes

Magnetic tape was the main storage medium in databases in the 1960s and early 1970s, but after this time the increased capacity and reduced cost of magnetic disks made disks almost the exclusive medium for databases until the arrival of the optical disk. The main reason that magnetic tape is no longer used is that to access a record at the end of a tape requires the whole tape to be read until the record is found, which may take 10 minutes or more. Thus tapes can only be used efficiently for sequential processing. Disks can also be used for sequential processing but in addition they can combine sequential processing with random access (i.e. direct access to any block on the disk) making the disk far more powerful. So tapes have now been almost completely replaced by disks for holding the main data in a database system. (The exceptions are very large databases with sizes in tens of gigabytes, e.g. seismic data of an oil field or data from satellite observations[15].)

Nevertheless tapes still have two important roles to play: archiving and keeping security copies of files, and sending data from one installation to another. Both roles may eventually be taken over by the optical disk.

Back-up tapes

Most databases need an archive to store old files which will rarely be needed in the future, releasing space on the disk for more active files. All databases also need back-up copies of files to be taken in case of a disastrous disk crash or failure leading to the loss of all or part of the data on a disk drive. Although such disk failures are now uncommon, one of us (FJS) has recently lost the whole of a technical database because of such a disk crash, on a disk provided by one of the leading manufacturers. Fortunately a back-up copy of the database had been taken the previous month; so only a few weeks work was lost. The frequency of such back-ups should obviously depend on the number of changes made to the database. A weekly back-up should be normal, late on a Friday or at the weekend when the system is not busy. A monthly back-up risks weeks of work being lost.

When the database is small, less than about 10 megabytes, back-ups are conveniently dumped to floppy disks. When the database gets much larger greater than about 10 megabytes, the loading and unloading of (say) five or more disks is tedious. Therefore, for any database above about 10 megabytes a tape streamer should be used to allow convenient and frequent copies of files to be taken. If tape (or the alternative, an optical disk) is not available, the tedious job of dumping to a large number of floppy disks will inevitably cause operators to delay such dumps, possibly leading to a disaster. The cost of tapes drives has been falling in price, like disks, and you can purchase one for a small computer for no more than a few hundred dollars. A few back-up systems use cassette tapes, video tapes or more recently helical scan or digital audio tape (DAT), which permit the recording of data at high density (2 gigabytes on one cassette tape). Most use standard $^1/2''$ tape to give greater reliability and compatibility, but a few use high-quality cassette tapes.

If the database is very large, say a gigabyte or more, it may be very slow to dump the data unto tape, e.g. if it takes 15 min to dump 30 megabytes to a tape, then one gigabyte would take 33 tapes and almost nine hours to dump! In this case, helical scan tape would be faster, but as a better alternative a large database should have special software to dump only those files which have changed since the last dump, or alternatively it should have extra large disks to hold a complete extra copy of the whole database. Even for smaller databases when there is a high priority for a continuous user service, with the minimum of downtime (e.g. a blood databank) a second or third disk must be maintained (possibly one of which might be optical) with updated copies of the database. However, tapes should also be available for back-up because a disastrous loss of both copies of the database is still possible due to a system software problem.

Communication

Tapes are a convenient method of sending data or programs between installations which have compatible tape decks. For small amounts of data or programs, floppy disks are often more convenient, provided that the two installations have compatible disk drives, e.g. IBM-compatible disk drives. Of course, programs or data can also be sent between installation over telephone circuits or data circuits, but until reliable digital communication, at a lower cost, is provided by the communications companies, postage of tapes, floppy disks or optical disks will remain more reliable and a lot less expensive.

7.1.4 Addressing data on a disk

As we have stated, the data on a disk are held on blocks with typically 512 – 2048 bytes per block. The blocks are stored on circular tracks and all of the blocks on one track can be accessed without movement of the read/write head, i.e. with an average access time of about 10 ms. If several disks are stacked on the one spindle with read/write heads able to read any of the tracks above or below one another, then this collection of tracks is called a cylinder. There will be one cylinder corresponding to each read/write head position and this will normally be numbered $C = 0,1,2...$ up to the limit on the number of cylinders. The tracks within the cylinder, each corresponding to a surface, will also be given a number, $T = 0,1,2...$ and the blocks on each track are likewise numbered, $B = 0,1,2...$ Words, bytes or sometimes bits can then be numbered within each block; so a data item might be located at

Cylinder $C = 5$
Track $T = 14$
Block $B = 227$
Byte $Y = 109$

with length (say) 12 bytes.

Some systems allow this definition of the physical address on the disks of each block; but to make the user interface easier, most systems allow you to define a file of N blocks which you can address with their numbers, between 1 and N, called **relative** or **logical addresses**. The blocks in the file are normally stored together on one track or cylinder; but they may not be. You may have appended data to your file at different times and the operating system or database may have stored each addition at a different part of the disk, linked by pointers, but you need not know this. You address the blocks as if they are all together; the system automatically maps each of your relative addresses (i.e. your view of the data) to the actual physical addresses (C,T,B,Y) on the surface.

However, almost all database systems facilitate even easier access to your data by allowing you to address files and records rather than blocks. You do not need to be aware of the blocks' structure underlying your view of the database as a collection of files of records; also, addressing a record in a named file is much easier than addressing a position, even a relative one, on the disk. We discuss structures for storing records in the next section.

7.2 RECORDS AND FIELDS

Most data in both business and technical databases can be structured in records, each record made up of a number of fields. This structure can be conveniently represented by a table and examples are given in Table 7.1 and Table 7.2. An important exception is text data or documents; we discuss these later in Chapter 9. For many scientific or engineering databases the diagrams, computer code, rules and formulae make an important part of the total knowledge which must be stored in the database. We discuss these at the end of this chapter (Section 7.6), but the main part of this chapter concerns the record-field format which is suited for the great majority of data. A collection of records which have the same fields and formats, similar to those in Table 7.1 or Table 7.2, is called a file (also sometimes called a dataset). The file

Table 7.1 Example of simple atomic mass records, each with four fields.

ATOMIC MASS FILE

	Atom	Symbol	Atomic Number	Atomic Weight
Record 1	Hydrogen	H	1	1.008
Record 2	Helium	He	2	4.003
Record 3	Beryllium	Be	3	9.015
Record 4	Oxygen	O	8	16.000

is always given a name, e.g. the file names corresponding to the records in Tables 7.1 and 7.2 will be the Atomic-Mass file and the Ferrous Alloy file.

Many computer languages have special features to make the handling of records and fields easier. For example, PASCAL has very useful structures for manipulating records provided they are stored in sequence [5]. Unfortunately there are no structures for random access to a disk store in standard PASCAL, although most implementations of PASCAL include such features. Most scientists and engineers use FORTRAN and fortunately there are special structures in FORTRAN for storing records either sequentially or randomly. In the example of atomic mass records in Table 7.1, if the records are stored sequentially then each record could be written to a file (unit 10) with statements of the form

```
      WRITE(10, 100) ATOM, SYMBOL, ATOM-NO, ATOM-WT
  100  FORMAT (A20, 3X, A3, 2X, I3, 2X, E7.3)
```

In this the format lengths are fixed at 20 characters for the atom name field, after

Table 7.2 Example of a slightly more complicated record on ferrous alloys with eight fields. Records in science and engineering often need extended fields to record additional information, as in record 5. Sometimes comments need to be added.

FERROUS ALLOY RECORDS

	Alloy	Form	Condition	Sigma (%)	Temp (°F)	Creep (%)	Time (hr)	Stress (ksi)
Record 1	310S	0.050" Sheet	2000 °F hr AC	0	1200	1	10	24
Record 2	310S	0.050" Sheet	2000 °F hr AC	0	1200	1	100	22
Record 3	310S	0.050" Sheet	2000 °F hr AC	0	1200	1	1000	20
Record 4	310S	0.050" Sheet	2000 °F hr AC	0	1800	1	10	5.3
Record 5	310S	0.050" Sheet	2000 °F hr AC + 1600 °F 200 hr	2.5	1500	1	10	10

three spaces three characters are given to the symbol name, after two spaces an integer of three numerals is given to the atomic number and after two spaces a seven-character field is allowed for the real number representing the atomic weight (three decimal places). Of course, to write all of the records to a file the write statement would have to be stored within a loop.

To retrieve from this file you use a similar read instruction within a loop; for example, to print the atom name with the atomic number of 92, (i.e. Uranium) we would write

```
99   READ (10,100) ATOM, SYMBOL, ATOM-NO, ATOM-WT
     IF ATOM-NO.NE.92 GO TO 99
     PRINT *, 'ATOM IS', ATOM
```

If these records were stored in order of the atomic number then the system would have to read 92 records before it reached the one needed, which obviously wastes a lot of processing time. This is fast enough when there are only 92 records to be read, but it would be much too slow if the search was through a catalogue of materials with 92,000 records to be read. If the file is on a tape you would have no choice, but on a disk file random access, i.e. direct access to the block holding the data needed, is much faster. Most programming languages have a structure for such direct access to a record. You can open a new direct file ATOMIC-MASS as follows

```
OPEN (UNIT = 19, FILE = 'ATOMIC-MASS', ACCESS = 'DIRECT',
RECL = 40)
```

The variable RECL is the record length. The records within such a file are numbered 1,2,3, etc. and are stored in file ATOMIC-MASS on UNIT 19. To store a record in the file with a record number equal to the atomic number we can write

```
WRITE (19, REC = ATOM-NO) ATOM, SYMBOL, ATOM-NO, ATOM-WT
```

where REC is the record number.

Usually records and direct files are unformatted. To print the element name for the atomic number 92 we would now use the following READ and Print statements:

```
READ (19, REC = 92) ATOM, SYMBOL, ATOM-NO, ATOM-WT
Print*, ' ATOM ', REC, ' IS ', ATOM
```

This will print

```
ATOM 92 IS Uranium
```

Details can be found in any text on FORTRAN 77 [6].

Note that in the above example we were able to set the record number = 92 because we knew that the record number corresponded to the atomic number. Normally these numbers would not correspond and you would first have to look up an index to find the correct record number. For a large file of records this index might be a B-tree (to be described later in Section 7.3.4); for a short file like this it could consist of a simple look-up table.

7.2.1 *Fixed length records*

In most files and in the above examples the records are of a fixed length, that is, there are a fixed number of fields in each record and each of its fields has a fixed length, usually measured in bytes, each byte being 8 bits in length. All of the records in the file will then have the same fields, each with the same predefined length and, of course, the total length of records in the file will therefore be the same. This is the most common type of record, much preferred by those who write databases, as the software needed for the processing of a file of fixed length records is obviously easier to write than the software for processing a file where the record length can change. In particular, it makes it easier to know that a fixed number of records are stored in each block on disk when the length of each record is known. Then, for example if there are 10 records per block we know that Record 73 is in block 8.

The disadvantage is that the predefined number of fields per record and the fixed lengths of the fields limits the information that can be contained within each record. For example, an author's surname field may be reasonably limited to 12 characters, which is sufficient for well over 99% of surnames. But Mr Willoughby-Smith may not appreciate his name being cut to Mr Willoughby-S and being confused with Mr Willoughby-Simms! By this truncation after 12 letters some information is unfortunately lost. A more serious example is illustrated in the fifth Ferrous Alloy record in Table 7.2. If we had restricted the length of the 'Condition' field to 15 characters, we would have been unable to include the vital additional condition of a delay of 200 hours at 1600 °F.

Such occasional loss of information is inevitable in many files of fixed length records and this fixed structure, though more efficient in computer processing time, must therefore be examined carefully and avoided when necessary in the planning and design stage of the database. Information loss is a problem in any database, either business or scientific, but it is more serious in science or engineering.

We have already given an example where a restriction on field length would result in a vital loss of information in an experiment. A restriction on the number of fields can also be serious. For example, a record may include fields for up to 10 measurements of each property. But because of a large variance in some measurement a scientist may decide to make many more than 10 measurements and then find the database cannot accept them because of the rigid structure of only 10 measurement fields. Information, perhaps never to be recovered, is then lost and less accurate information (e.g. on the best average value for the property) or erroneous information (e.g. leaving out the delay of 200 hours at 1600 °F) is then given to all future users. This is an example of man being restricted and hindered by the database tool rather than being served by it.

One way we can avoid the loss of information in a fixed length environment is to make every field a little longer than we would ever expect it to be, and to include more fields than the maximum number we think likely. Even if we do not underestimate these, by allowing for the largest possible record, much space is wasted when storing the shorter records. But in addition to this, the number of records per block is reduced and any index to the file is larger and the number of disk accesses needed to find a record is increased.

7.2.2 *Variable length records*

Variable length records use available storage space economically and also overcome the problem of occasional information loss in fixed length files. However, they require more software than fixed length records and more computer time to process them. A variable length record may have a variable number of fields and the field lengths may also vary.

Normally the record begins with a word, called the record descriptor, recording the length of the record, including all of its fields and the word descriptor itself. Each of the fields, if variable length, may itself be preceded by a descriptor giving its length in bytes. Alternatively a field delimiter, i.e. some character which cannot occur in a field, is used to separate fields. A comma may be a convenient delimiter to separate fields, as this is used automatically as a delimiter in some computer languages, e.g. FORTRAN and BASIC.

In Tables 7.3 and 7.4 we give two variable length record formats of the sample file on atomic mass in Table 7.1. In Table 7.3 a record delimiter is used to separate records rather than a record descriptor giving its length; a convenient delimiter is a carriage return and line feed, as they automatically put each record on a new line, making the file easy for us to read and for a computer program to process. If the field delimiter is chosen to be a comma, this also makes it a good human as well as computer interface (particularly for FORTRAN).

Variable length records, like fixed length records, are stored in blocks, but the number of records per block will now vary. With fixed length records, as explained earlier, we could quickly compute the block on disk holding a record. For example, if a record is 50 bytes long and a block has 512 bytes then we would store 10 records per block. So record 24 would be the fourth record on block 3. Such a calculation is no longer possible with variable length records. We need an index or table to tell us where records are stored.

However, normally when we search for a record we do this by specifying some

Table 7.3 A variable length format for the records in Table 7.1 using a comma as a delimiter between fields and a carriage return and line feed as the delimiter between records. In this form it can be easily processed by a FORTRAN program.

Hydrogen, H, 1, 1.008

Helium, He, 2, 4.003

Beryllium, Be, 4, 9.015

Oxygen, O, 8, 16.000

Table 7.4 A variable length record format for the records in Table 7.3 using record and field descriptors. The number 8 is the length of a real number (e.g. 1.008 EO). Each integer takes 2 bytes.

```
28/8/HYDROGEN/1/H/1/1/8/1.008E O/
27/6/HELIUM/2/He/1/2/8/4.003E O/
30/9/BERYLLIUM/2/Be/1/4/8/9.015E O/
26/6/OXYGEN/1/0/1/8/8/1.6000E 1/
```

---

key or data item, e.g. we might search for the atom 'Uranium'. The system then uses a look-up table or index to find first the record number and then from this the block number, i.e. it may find that the Uranium record is on block 24 (say). This allows rapid access to the record. We discuss different forms of index in more detail in Section 7.3.

7.2.3 Variant records

Variant records allow the fields of variable length to be extended further to allow records composed of different fields to be included in one file. For example, we might have an Atomic Particle Mass record which would be similar to our Atomic-Mass record in Table 7.1 but which would include the interatomic distances and molecular weight as fields when the particle is a molecule and the atomic number and atomic weight when the particle is an atom, e.g.

```
Record 23: 1, Hydrogen, H, 1, 1.008
Record 39: 2, Hydrogen Chloride, HCl, 1.042, 36.465.
```

The first number in each of these records is now the number of atoms in the molecule.

In FORTRAN this would be read with conditional statements such as

```
READ (10) NO-OF-ATOMS
IF NO-OF-ATOMS = 1 THEN READ (10) ATOM, SYMBOL, ATOM-NO, ATOM-WT
   ELSE READ (10) MOLECULE, SYMBOL, ATOM-DISTANCE, MOL-WT
END IF
```

But, in general, we do not recommend such a structure. In the example above it would be better to have two files, one for atoms and a second for molecules.

7.3 FILE ORGANIZATION

We have already explained in Section 7.1 that the performance of a database is determined primarily by the number of disk accesses required to perform each task. So it is vitally important, for good performance, to provide file structures or file organizations to minimize these disk accesses.

In many database building situations, the builder will use commercial software and will not have to choose and implement any of the methods described in this section. However, often the performance of databases built with commercial software is very disappointing and understanding the different file organization techniques will allow builders to understand what is wrong and to take steps if possible to improve this performance.

The organization methods described are

1. serial files
2. sequential files
3. indexed sequential files
4. B-trees
5. hierarchical trees
6. linked lists (chains and rings)
7. hashed files and
8. inverted files.

The first of these are simpler methods of organization than the later more complex methods. The complexity arises particularly because some databases, called active databases, have frequent additions, insertions, deletions and corrections that make any simple natural ordering change with time. Active databases therefore require more complex structures than static databases which do not change.

7.3.1 Serial file

If the processing of records is always serial, one after the other, and if the order of processing the records is not important, then the most convenient file structure is a serial file where the records follow one another in any order and where each new record added is put at the end of the file. Records are kept adjacent to one another to speed processing.

However, it is unlikely that this very simple structure will be satisfactory for many applications, because if you want to find a particular record in the file then you have no choice but to read through the whole file to find the record. Such a serial search may be very slow if the file is large. So one of the following structures, which make random access faster, is then more appropriate.

7.3.2 Sequential file

The most common ordering of a file is sequential: records are sorted and stored in sequential order of the important key values in one or more of the fields, i.e. in alphabetical order of a name field. For example, the file in Table 7.1 of atomic masses is ordered on the atomic number. If it was ordered on the name of the atom then it would start with Actinium, as illustrated in Table 7.5.

Records in sequential order (like all records on disk) are held in blocks. In the case of a static database when we know that no new insertions will be made, the blocks

Table 7.5 Part of the Atomic-Mass file, ordered alphabetically on the name of the atom.

Atom	Symbol	Atomic Number	Atomic Weight
Actinium	Ac	89	227.0
Aluminium	Al	13	26.97
Antimony	Sb	51	121.76
Arsenic	As	33	74.91

can be filled. However, in an active database, to facilitate future insertions part of each block is left unfilled (typically 30%). If this is not done, a subsequent insertion at the beginning of the file will require all the later records to be moved forward one place. If space is left within the block where the insertion is to be made, then only that one block has to be reordered and not the rest of the file.

Once the records are in sequential order it is much easier to find records corresponding to a particular value of the sorted field. For example, in the Atomic-Mass file in Table 7.5 if you wish to find the record for the atom 'Uranium', you could perform a binary search as follows. The block in the centre of the file is examined. Since the search key is lower in the alphabet than the last record in the block, we next examine the block in the centre of the bottom half of the file and continue this process until a block containing the record is found (in this case the last block).

Binary tree

If there are N records in the file and an average of m records per block then the number of disk accesses, as a_2, needed to find a record using a binary tree is approximately

$$a_2 = \log_2(N/m).$$

This is much better than a serial search where an average number of disk accesses is

$$a_1 = N/2m.$$

However, if we used a tree structure (see Section 7.3.4) with several branches, r, from each node of the tree, rather than 2, the number of disk accesses would be approximately

$$a_r = \log_r(N/m),$$

substantially less than a_2 if r was large. For example, if $N/m = 1000$ then

$$a_1 = 500$$
$$a_2 = 10$$
$$a_{10} = 3.$$

The next two file structures, indexed sequential and B-tree, are examples of tree structures with several branches.

7.3.3 Indexed sequential file

When a separate index is made for one or more of the fields of a sequential file it is called an indexed sequential file. An example is illustrated in Figure 7.2. The first level index is the set of the highest values of the key field within each block of the main file and the second level index is the set of the highest values in the first level. For large files several levels are needed to index all of the records. This structure is then equivalent to an n-ary tree where there are n keys in each block of the index. This number n should be kept lower than the capacity of each index block to facilitate future insertions of records to the main file and possible subsequent insertions in the index file if the number of blocks in the main file.

However, it presents a difficulty if there are many insertions. Even if space is left in each block, they may frequently be over-filled and spill over to the next block. Since the highest record for each block will be the first to spill over to the next, the first level index and possibly higher levels will have to be changed.

To avoid this, if sequential processing is not frequent, additions to any block, A, in the main file are allowed to spill over to an additional empty block, A' often at the end of the file, and a pointer inserted in A linking it to A' and another in A' linking it back to the block, B, which previously followed A. The resulting structure is illustrated in Figure 7.3 for the insertion of the Author names 'Ashford' and 'Aston' in the example in Figure 7.2. After resorting the contents of the block, two Author records are moved to the end of the file and the index adjusted accordingly. Note that the insertion of Author records for Bell and Wilson have been made to the Author file without any changes outside blocks 2 and 15 because space was available in both these blocks for insertions.

A multi-level index as in Figures 7.2 and 7.3 take the form of a tree structure and when the tree is balanced we call it a B-tree. This is explained in the next section.

7.3.4 B-tree

The most common method of indexing a file in database management systems is the B-tree, or Balanced-tree structure [7]. An example of a general tree is illustrated in Figure 7.4. At the top of the tree is the root-node and all other nodes are joined to the root node by a sequence of branches. Each node can have only one parent node. The nodes at the bottom of the branching structure, with no child nodes, are called leaf nodes or leaves. For example, in Figure 7.4 the leaves are marked L1 to L6. These leaves represent the records which are being indexed.

When the number of branches from the leaf nodes to the root node is the same for all nodes, the tree is said to be a balanced tree. If this is used for indexing then the number of steps needed to find each record is uniform for all records. This is the basis of a B-tree.

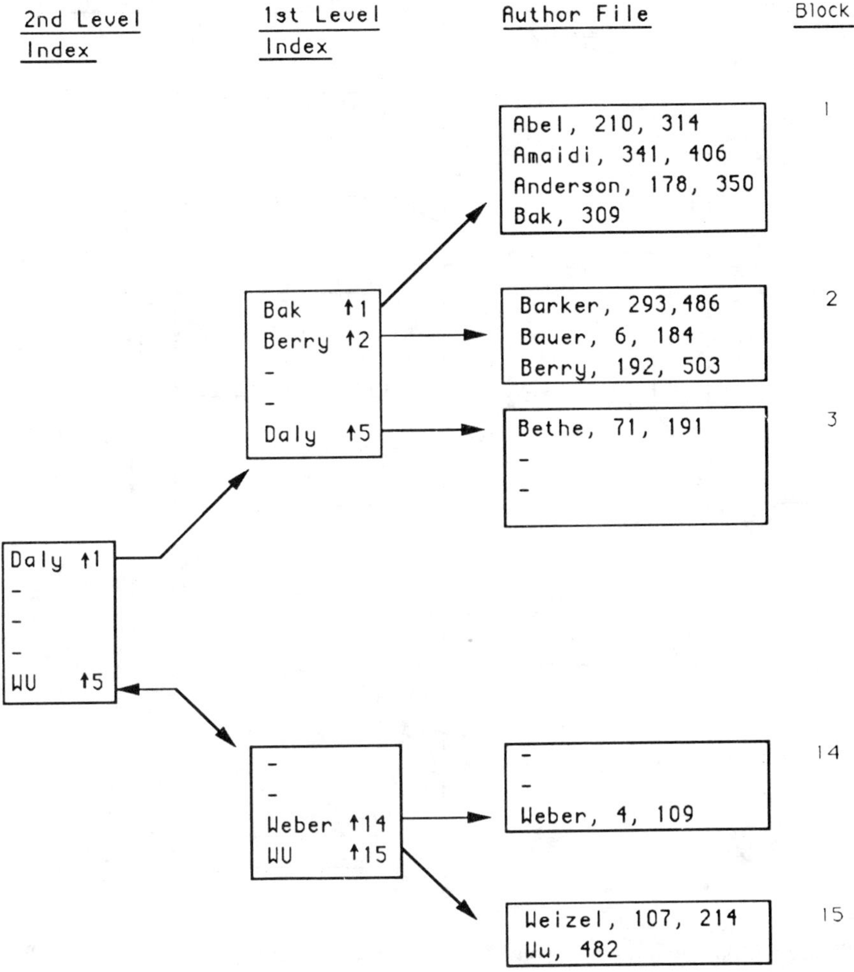

Figure 7.2 Example of an index to a file of authors' names and references. If the index is long a second level index to the first level of the index can be provided and so on, producing a tree with many branches (Note: The last author name, Wu, need not be included in the index.)

The structure is the same as the structure of an indexed sequential file in the last section; however, it is built and modified as the database changes using a few well-defined algorithms which can be applied to any sequential data and which maintain a tree structure which always balances, irrespective of the number of insertions and deletions.

An example is given in Figure 7.6. This illustrates that the records in the main file are no longer in sequential order but are indexed by unique keys identifying the records at the bottom of the index stored along with pointers to the records. This

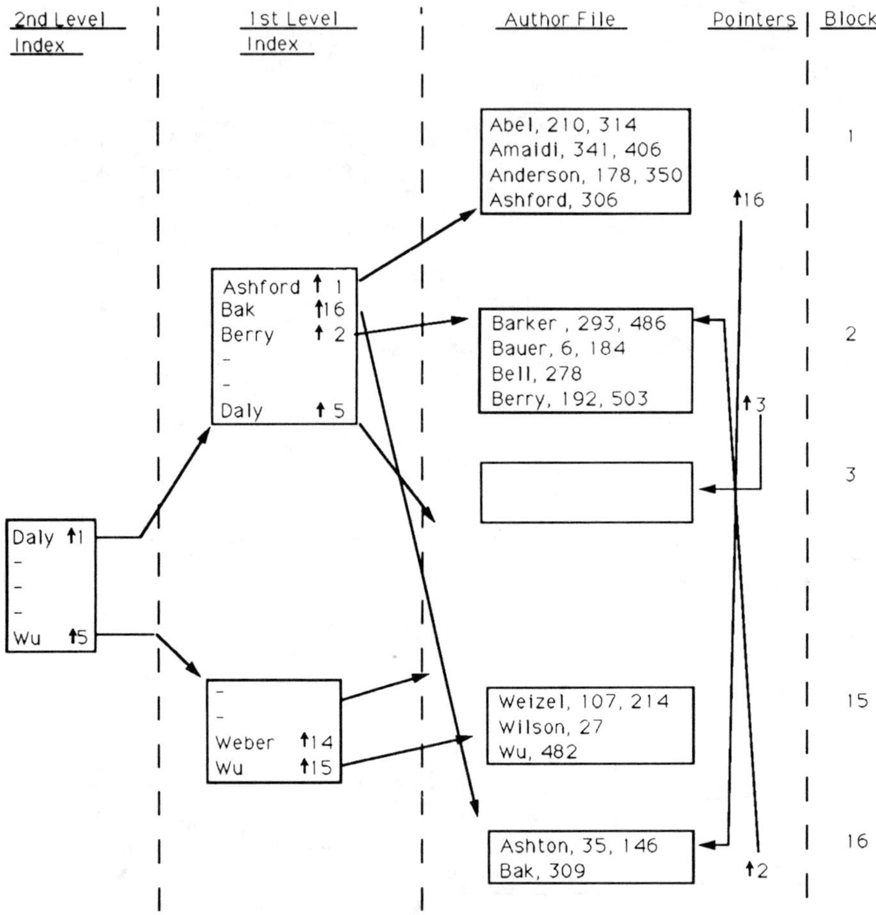

Figure 7.3 Illustration of use of pointers to allow insertions of records (Ashford and Ashton) to block 1, causing overflow to block 16 at the end of the file (assuming a capacity of four records per block). The file does not have to be reordered, only the index holding the highest key value in each block needs to be reordered. Note that records for Bell and Wilson have been added without changing the index.

bottom level to the index is often called the 'sequence set'. The multi-level index to the sequence set is called the index set. The order of the index is the maximum number of keys held at each node or block and is one less than the number of branches in the tree. The last branch at each node corresponds to a key value at one point higher in the tree (see Figure 7.4 for an example) or to the last key (which need not be stored).

  Searching for a record in a multi-level index (as in Section 7.3.3) or in a B-tree

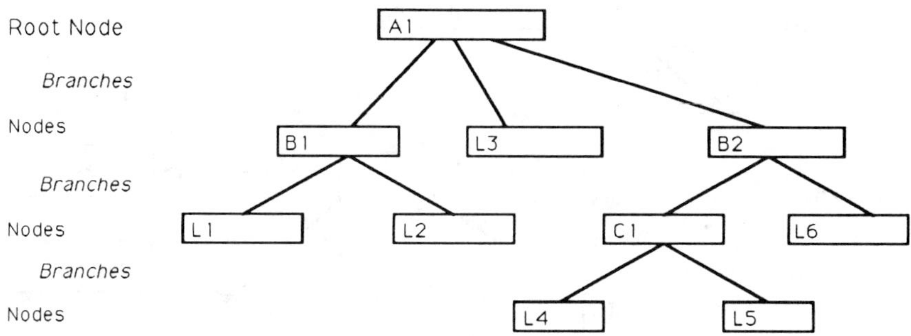

Figure 7.4 Example of a tree structure. Nodes L1 to L6 at the bottom of the branching structure are called leaf nodes.

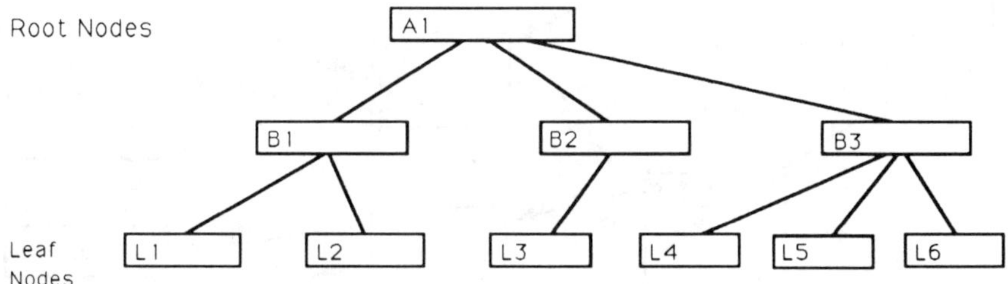

Figure 7.5 A balanced tree, in which the number of branches to all leaf nodes are equal.

is essentially the same simple process which we describe in the algorithm below.

[Comment: search for sequence set block containing X]

```
Let N be the top node of the index set;

Repeat
    Let X₁ < X₂ < .. < Xₘ be the key values in node N

    If X ≤ X₁ then follow the pointer above X₁ to a
    node at the next level of the index;

    If X > Xₘ then follow the pointer below Xₘ to a
    node at the next level of the index;

    If Xₖ < X ≤ Xₖ₊₁ then follow the pointer between to a
    node at the next level of the index;

    Let N be the new node;

    until N is a sequence set node;
```

One of the attractions of a B-tree is that a simple algorithm can be constructed for the insertion (or deletion) of key values in the index always keeping the tree balanced. We illustrate this with the two algorithms needed for the insertion of a new record with key X in the tree.

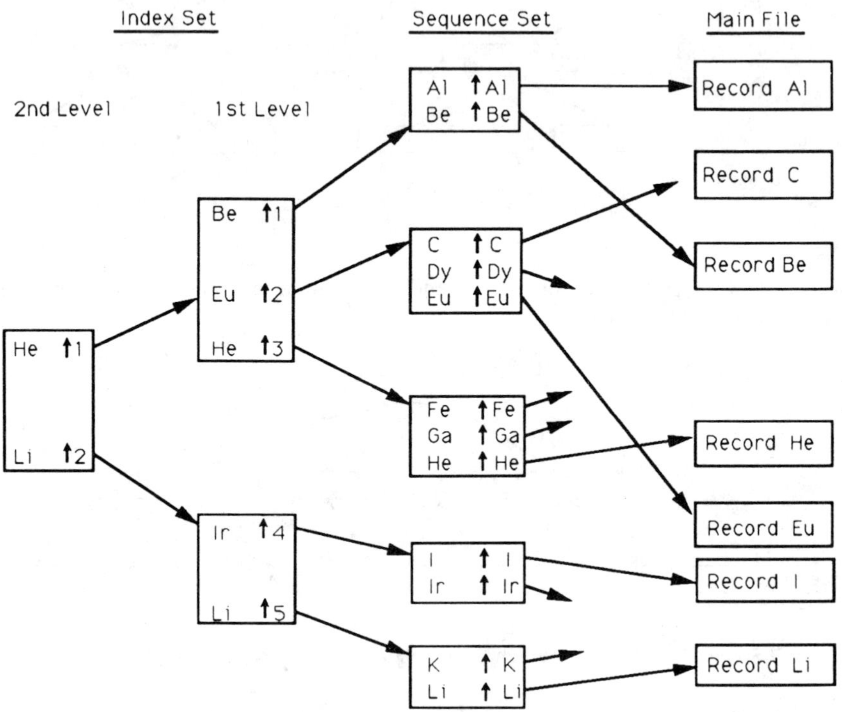

Figure 7.6 Example of a B-tree. Note that the records, stored in blocks on disk, are not necessarily ordered; only the index to the main file (the sequence set) is in order.

[Comment: insert X and a pointer ↑X to the record X in the sequence set]

```
Let X₁ < X₂ < X₃ ... < Xₘ be the key values in
node N;

If N is not full insert X and ↑X;

If N is full

    Sort X₁ X₂, ... Xₘ, X in order;

    Break into two equal or almost equal subsets L and H
    (including pointers) where L < H;

    Replace N by H;

    Create a new sequence set node L;

    Insert the highest key X' of L and its pointer
```

```
to L, ↑X', into the superior index set node; [using
the following algorithm]
```

[Comment: to insert X' and ↑X' in an index node, N]

```
Let X₁ < X₂ ... < Xₘ be the key values of N;

If N is not full insert X' and ↑X' a pointer to the next
inferior node corresponding to X';

If N is full

    Sort X₁, ... Xₘ,X' in order;

    Break into two equal or almost equal subsets L and H
    (including pointers) and let Xₖ' be the highest
    key of L;

    Replace N by H;

    Create a new index node L;

    Let X' = Xₖ';

    Insert X' and ↑X' in the superior node of the tree.
```

A B-tree formed by this algorithm is always balanced, i.e. the number of branches needed to reach the keys in the sequence set are always the same, ensuring that the search time, in disk accesses, is the same for all records.

Note that on average a B-tree may have only three-quarters of each index block or sequence block full. However, the number of keys and pointers in a block may be large, say 64 for a block of 1024 bytes. This allows about 32 branches at each node and a million records can be accessed with

$$a_{32} \, (1\,000\,000) = 4 \text{ disk accesses}$$

(plus one for the top node and final record block). Thus any record in a file of 1 million can be accessed with 6 disk accesses in total.

The convenience of a B-tree, the algorithms for insertion, deletion and searching and its speed make it probably the most common form of index used in databases. It is also easy to understand.

However, to obtain the highest possible performance at the price of being less user-friendly, you should use a hashed index. There are several forms to the B-tree, but all are the same in principle; so we need not discuss them here. They are described in detail by Semet [2].

7.3.5 *Hierarchical tree*

A B-tree is a structure for storing and indexing a sequential set of records which usually do not have any naturally occurring hierarchical structures. When one exists, then a hierarchical tree structure can be used.

Much scientific and engineering data are naturally structured in the form of a tree. For example, the Ferrous Alloy records in Table 7.1 can be suitably (perhaps more suitably) stored in the form in Figure 7.7. This structure is the basis of the Hierarchical and Network models of a database which we discuss in Chapter 9; so we do not need to discuss it here further.

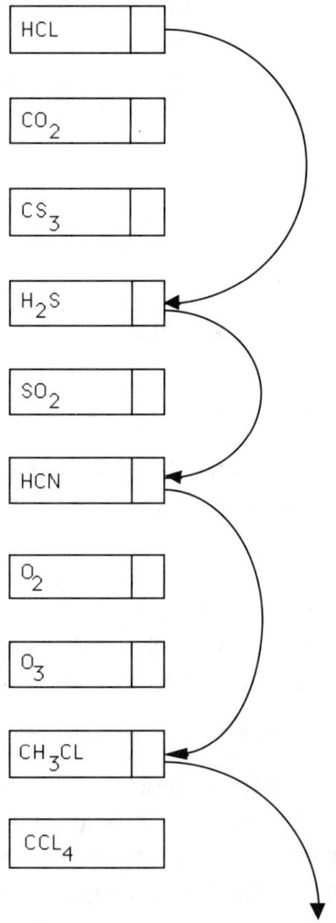

Figure 7.7 Example of the use of pointers to provide a chained index to molecular records containing Hydrogen because of a need for frequent access to Hydrogen molecules.

7.3.6 *Chains and rings*

Records linked together with pointers so that each record points to the next are called chains or linked lists. When the last record points back to the first the structure is called a ring (or sometimes a loop). Since these structures allow records to be situated in any physical order on the disk, they are likely to involve many disk accesses before processing all of the records in the chain or ring. So they should only be used when a faster alternative is not possible.

They are commonly used to store lists of related records which have many different numbers of records per list. An example is the list of occurrences of words in this book. The word 'the' would occur thousands of times, the word 'database' possibly a hundred or more times, but the word 'loop' occurs only four times. When the length varies as much as this, a chain structure is a convenient representation. Such a structure is used in the Inverted List model of a database and in text database management systems (see Sections 8.4 and 9.5 of the next chapter).

A chain or ring is also convenient for secondary indexing. Consider a file concerning properties of molecules in a hashed order, or some sequential order. Each molecule is made up of several atoms. If we needed to process frequently the records for the molecules containing Hydrogen, then each molecular record including Hydrogen could include an additional pointer linking it to the next molecule including Hydrogen. It is illustrated in Figure 7.8.

A secondary index of this form can be very useful; however, it should be noted that it will still take many disk accesses to answer a query as the system has to jump from record to record on the disk—but normally this will be much less than the alternative, reading the file serially from beginning to end.

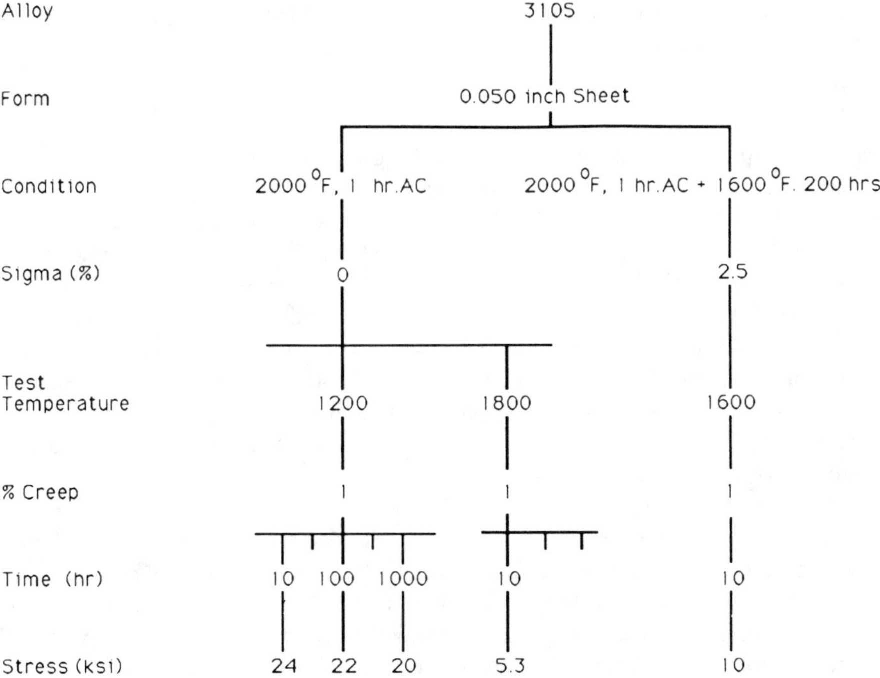

Figure 7.8 Representation of the data in Table 7.2 in a tree structure.

7.3.7 Hashed file

A hashed file (sometimes called a direct file) is one which uses an algorithm to compute the address from the key value of a record, or rather from some numeric value dependent on the key when the key is not a number. Many algorithms can be used [8], but one of the most common is known as the division hash algorithm and we describe it below.

Division Hashing

Let the file storing the records be held in B blocks on the disk, numbered 0, 1, 2 ..., $B-1$; then the hash address of the record with key value, K, will be

$$H_O = K \bmod B$$

which is the mathematical way of writing the remainder, after dividing K by B. If the key is an alphanumeric name (or code) rather than a number, then a numeric key, K, can be computed from the ordinal value (numerical equivalents in the binary code) of the characters in the name, using a formula such as

$$K = \mathrm{ord}(C_1) + 256*\mathrm{ord}(C_2) + (256)^2*\mathrm{ord}(C_3) + (256)^3*\mathrm{ord}(C_4)$$

where C_1, C_2, C_3 and C_4 are the first four characters of the name. To separate names with the same first four characters, C_5 can first be added to C_1, C_6 to C_2 and so on before computing K.

Overflow

The main problem with this form of hashing is that overflow of blocks can be common; this is the situation which occurs when our algorithm generates the same address for $C+1$ or more records where C is the capacity of a block to hold records. Then the surplus records must be stored somewhere else. This happens much more often than is intuitively obvious, particularly when the capacity of a block is small and when the file size is large. For example, in a file which can hold 10000 records, stored one per block, a simple statistical argument shows that two records will probably fall in one block, causing overflow, after only 100 records are stored (this is a form of the well known birthday paradox that two people in a room of 27 people are likely to have the same birthday). So surprisingly overflow starts in this case with only 1% of the file filled.

Because of the large number of chance clashes in large files, division hashing was not recommended until recently. It is now understood that the overflow problem is very much less serious if the number of records per block is large and if the packing density, that is the fraction of the file which is filled, is kept less than about 75%.

The method of storing the extra records which overflow is also vital. A special area for storing overflow records is not recommended—it might overflow itself long before a large file is 75% full. A simple algorithm for storing overflow records is to attempt to store them in the next block, i.e. if block 5 is full we try block 6 and repeat in block 7, 8, 9, etc. until space is found to store the record. This is the same as the generation of a set of addresses H_i for $i = 0, 1, 2, 3$, etc where

$$H_i = (K + i) \bmod B$$

and K is the key and B is the number of blocks in the file. The first of these, H_0, is the remainder. Unfortunately, if (say) four blocks in sequence are full then for new records being added to the file at random, the fifth block will fill up at five times the rate of other blocks and is likely also to overflow. The sixth block will then overflow too, and a cascade will form growing exponentially in length. Long cascades are common in large files and they can badly affect the average access time, very much more than is intuitively obvious. So this method should be avoided.

One convenient solution is to use a formula for generating subsequent overflow addresses, H_i, which depends on the original numeric key value, K. Several of these are possible but one of the simplest, the linear quotient method, we have found to be very effective[10]. This depends on the quotient Q obtained after dividing K by B and takes the form

$$H_i = \begin{cases} (K + i*Q) \bmod B & \text{if } Q \bmod B \neq 0 \\ (K + i) \bmod B & \text{if } Q \bmod B = 0. \end{cases}$$

By using this method of storing overflow records, by keeping the packing density less than 75% and by storing several records per block (> 10), the average number of disk accesses needed to find a record can be less than 1.1 accesses, even for large files. In general, the higher the packing density and the smaller the number of records per block, the more overflows. In particular the situation where one record is stored in one block should be avoided.

If records are large, than a separate index for all records can first be constructed, similar to the sequence set used in the construction of a B-tree. Since the basic unit of an index consists of a key value and a pointer; so a large number can be stored per block. This keeps down the overflow problem.

Within such a block the keys can be stored in numeric or alphabetic order; but it is usually more efficient to store them in order of the probability of search, i.e. those sought most often are stored at the top of a block.

Extendible hashing

The problem with division hashing is that the physical size of the file holding the records is fixed. If the packing density of the file increases above about 80% the amount of overflow increases rapidly and the response times lengthen. When we try to store more records than the fixed file space can hold, the system crashes unless a larger fixed file space is allocated. Since a larger divisor is needed to distribute the records over the bigger file, a new hashing algorithm has to be applied to the keys of all of the records in the file and many of the records, perhaps most records, have to be moved to their new positions. This is a very long procedure.

An alternative is some form of extendible hashing which allows only an overflowing block to be redistributed over a larger number of blocks. The principle was not new; an early method was used effectively by Higgins and Smith in 1971 [11]. However, it was not until 1979 that a fast economical method was devised by Fagin *et al* [12].

The method is illustrated in Figure 7.9 a hashing function, possibly a division hash with a large divisor is applied to the key for each record, K, to complete a pseudo-key K'

$$K' = h(K).$$

The first d bits of the pseudo-key are used to form a directory with 2^d pointers to the blocks holding the records. The value d is called the depth of the index. An important feature of the method is that the number of blocks in the file is normally much less than the number of points because often two or four or more pointers are aimed at one block. This is illustrated in Figure 7.9.

When a block overflows with the addition of a new record, the records are distributed over two new blocks. If two pointers are aimed at the block which overflows, say the block corresponding to K' starting with 01, then the records are distributed so that the two new blocks correspond to pseudo-keys starting with 010 and 011 (i.e. by adding another binary bit). The two pointers now point to these two blocks. However, when only one pointer is aimed at the block which overflows, another bit has to be added to the address and the depth d has to be increased by one.

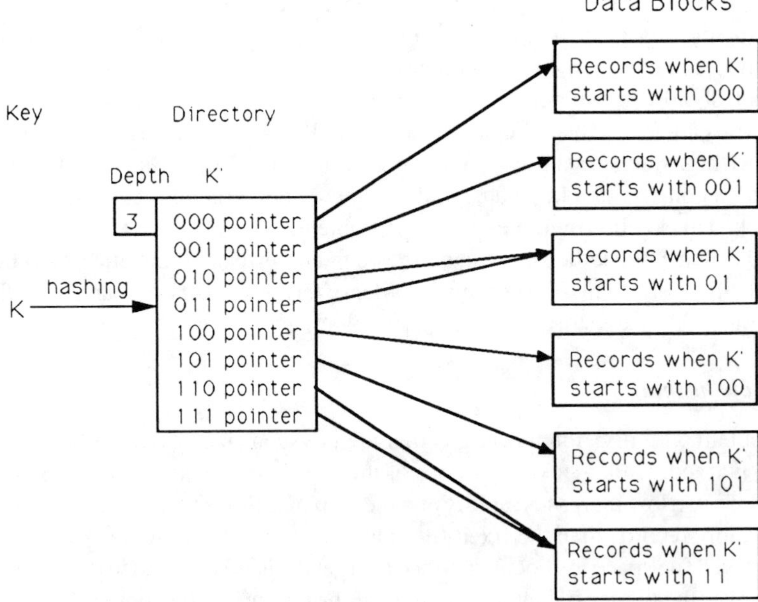

Figure 7.9 Illustration of extendible hashing for depth d=3. K' is a pseudo-key obtained by hashing the key K.

This requires the doubling of the size of the directory, but this is a great deal simpler than doubling the size of the whole file.

Extendible hashing requires two disk accesses to retrieve a record, one to find the pointer in the directory and a second to find the block holding the record. This is faster than a B-tree for large files, but slower than the division hash algorithm when the packing density is less than about 75%. However, for a very active database, changing rapidly, the extendibility of the file is a major advantage and extendible hashing is usually preferred to division hashing.

A large number of variations on extendible hashing have appeared in the literature since 1979 [2]. A brief and useful overview of the techniques for an engineer or scientist is given by Elmasri and Nayathe [13].

7.3.8 *Inverted file*

If a separate index is provided for every field then the file is said to be inverted. For example, in the case of the Atomic-Mass file in Figure 7.2 if an index is provided for the atom name, the symbol, the atomic number and atomic weight, then the file is inverted. Note that it should be possible to reconstruct the original file from the accumulation of inverted files.

However, the term 'inverted' is sometimes used to mean 'indexed' on only one field, e.g. the phrase 'inverted on field X', means that a separate index has been created for field X.

One of the most commonly used inverted files is the one for free text where an index is made for every word in the text. Most text databases (information retrieval systems) are based on this principle. Text databases are discussed in the next chapter.

7.4 DIAGRAMS, IMAGES, CONTINUOUS SIGNALS

In science and engineering, data in the form of diagrams, figures or photographs or measurements of continuous signals such as spectograms are an integral and vital part of the information which must be stored and later retrieved. Unfortunately few database systems provide facilities to store such information and the scientist usually has to build a special system to deal with them.

When a diagram is simple enough that it can be represented by a graph or by a reasonably small number of curve plots, then the data can be stored in the form of tables, possibly a long table in the case of a spectogram, but easily represented as a file of records. However, the actual photograph of a spectogram or a complex circuit diagram is often best stored as a digitized image. An optical disk (see Section 7.1.2) is suitable for digital recording of images (tens of thousands can be stored on one disk); but remember that a filing cabinet can also hold a lot of photographs at a low cost. In both these cases if the images are numbered or labelled, the numbers or labels can then be stored easily in any conventional database.

7.5 CONTINUOUS VARIABLES

Much data in science and engineering represent values of properties which may vary continuously with independent variables such as temperature or pressure and which represent properties or conditions of a substance or material. These continuous variables are usually stored as discrete values in fields describing the condition of the material. However, few databases distinguish between variables which record discrete values (such as atomic number or alloy number) from continuous variables (such as temperature). Unlike the example in the Atomic-Mass file in Table 7.1, which deals only with discrete quantities as independent variables, the example in Table 7.2 and in Figure 7.5 on properties of ferrous alloys deals with independent variables which are continuous. For example, in Table 7.2 stress varies continuously with the test temperature. Measurements may exist only at 1 000, 1 200, 1 400, 1 600 and 1 800 °F, and an unsuitable database management system (and in this context most database systems would be unsuitable) will respond with no answer if you ask for the stress at 1 300 °F; but we know that an accurate answer can be interpolated. You will normally need a special program to do this for you. There are no special data structures for continuous variables in general database systems on the market at present.

7.6 FORMULAE, SUBROUTINES, CODE AND RULES

A scientific textbook or an engineering handbook will normally include formulae giving relations between properties of the substances or materials. Regrettably such important information is usually left out of database systems because no structures are provided for storing such information, even though it is so vital. One reason is that it is not perceived to be needed in business computing. The inclusion of such information is the subject of much research, often under the heading of object-oriented databases (see Section 9.6). However, much can be achieved before this research produces working systems.

7.6.1 Formulae

The inclusion of a limited set of formulae in a database can be straightforward. One method is to include code for the computation of a general parametric formula in special subroutines and compile these into the software used to retrieve and manipulate data for the user. You then store the parameters in the database. For example, one parametric formula, often used, is a power series or exponential series. Single examples of these can be written in FORTRAN as follows:

```
      FUNCTION G(I, X, N, C)
      DIMENSION C(N)
      GO TO (1, 2, 3, 4, 5) I
    1 G = C(1)/X**C(2) + C(3)/X**C(4)
      RETURN
    2 G = C(1)*EXP (C(2)*X) + C(3)*EXP(C(4)*X)
      RETURN
```

For the parameter $I = 1$ this can represent functions of the form

$$Ax^{-n}, Ax^n, Ax^n + Bx^m$$

and for $I = 2$, all functions of the form

$$Ae^{-nx}, Ae^{-nx} + Be^{-mx}$$

and with $I = 3, 4$, etc we can define other general functions of the same type. A range of such functions can usually be made to cover all of the functions needed in a particular database and the parameters I, N, $C(1)$, $C(2)$, $C(3)$ etc can be stored as integers or real numbers in the database and the function executed every time the data are needed.

7.6.2 Rules

Rules which are usually simple formulae operating on 'facts' and probabilities can usually be stored in the same manner. The only difference is that they mainly work with Boolean variables which have two values, True or False. If you have to store and manipulate many rules and facts then an expert system or knowledge base system may be more appropriate than a database system. This is discussed later in Section 10.3.6 and at length in Chapter 12.

7.6.3 Implicit code

Code, or whole programs, can be packaged in the form of a subroutine and then handled implicitly in a similar manner to the parameterised functions in Section 7.6.1. The input data for the subroutine become the parameters held in the database and the subroutine is compiled into the program used to retrieve data from the database.

7.6.4 Explicit code

If there are too many formulae or too many separate pieces of code or programs or if they are changing too frequently to be compiled into your retrieval software it is possible to store code explicitly in the database in a form similar to a free text file along with other data. For example, let us assume we have a piece of FORTRAN source code which computes approximate extrapolations of the stress on an alloy for temperatures above the highest value measured. If you wish to obtain a value at a high temperature, your retrieval program can first note that the temperature is above the highest recorded value, then call down the source code into store, and suspend operation of itself while the source code is compiled and executed with the appropriate parameters (e.g. the last measured values of the stress). The result is stored, the retrieval program begins execution again and then displays or prints the approximate value. Unfortunately, no general database system will do this for you, but most comprehensive operating systems such as UNIX will allow you to write your own software for such a purpose.

It is also possible to store the semi-compiled code or binary code in the database. But then the code has to be linked to your user program. You will need the help of a systems programmer to do this.

However, fortunately in most cases, if you need to store formulae, you should try to use the first method, general parameterised formulae. One of us (FJS) had been using such a method for over a decade on several atomic database systems [14].

7.7   CONCLUSION

We have discussed almost all of the structures which are used for the representation of data in databases. Fortunately for most users, the database system hides the basic structures used in the database. The user may be able to 'INDEX' a field but not know which method of indexing is used (most likely, B-tree, see Section 7.3.4). However, it is always worth carrying out a test to check that whatever method is used is efficient. You should use a sample file, searching for the last record (i) when the file is short (say 10 records) and (ii) when the file is long (at least 1 000 records).

As we have explained above, the time to find a record should only increase logarithmically, i.e. very slowly with the size of file. Sometimes, however, you will find in a poorly designed database management system that the search in the long file takes many times longer. We know of cases where all so-called random access searches were actually serial and therefore very slow and wasteful of computer time and, more important, of your time. Avoid such database systems!

In the next chapter we discuss the general architecture of database systems which is needed before a detailed discussion of the principle data models used in database management systems.

REFERENCES

[1]   R. Elmasri and S. B. Nayathe, *Fundamentals of Database Systems*, Benjamin/Cummings Publ. Co., Redwood City, California, Chapters 4 and 5, 1989.

[2]   H. Samet, *The Design and Analysis of Spatial Data Structures*, Addison-Wesley, New York, 1990.

[3]   S. Apiki and H. Eglowstein, The Optical Option, Byte, 14, No. 10, 160-80, Oct. 1989; J. J. Burke and B. Ryan, Gigabytes On-Line, Byte, 14, No. 10, 259-65, Oct. 1989.

[4]   Crystal Data on CD-ROM, Office of Standard Reference Data, National Institute of Science and Technology, Gaithersburg, Md.;
Diffraction Data on CD-ROM, International Center for Diffraction Data, Swathmore, Pa.

[5]   J. Welsh and J. Elder, *Introduction to Pascal*, 3rd Ed., Prentice Hall, New York, 1988.

[6]   A. Balfour and D. H. Marwick, *Programming in Standard FORTRAN 77*, Heinemann Educ. Books, London, Chapter 19, 1979.

[7]   R. Elmasri and S. B. Nayathe, Fundamentals of Database Systems, Benjamin/Cummings Publ. Co., Redwood City, California, Section 5.3, 1989.

[8]   S. P. Ghosh, *Data Base Organization for Data Management*, Academic Press, New York, Chapter 4, 1977.

[9]   T. G. Lewis and C. R. Cook, Hashing for Dynamic and Static Internal Tables, *Computer*, **21**, 45–56, 1988.

[10]   J. R. Bell and C. H. Kaman, The Linear Quotient Hash Code, *Comm. ACM*, **13**, 675–7, 1970.

[11]   L. D. Higgins and F. J. Smith, Disc Access Algorithms, *Comp. J.*, **14**, 249–53, 1971.

[12]   R. Fagin, J. Nievergelt, N. Pippenger and H. Strong, Extendible Hashing - A fast Access Method for Dynamic Files, *ACM TODS*, **4**, No. 3, 1979.

[13]   R. Elmasri and S. B. Nayathe, Fundamentals of Database Systems, Benjamin/Cummings Publ. Co., Redwood City, California, Section 4.8.3, 1989.

[14]   J. Boyle, W. R. McDonough, H. O'Hara and F. J. Smith, An On-line Numerical Data System, *Program*, **11**, 35–41, 1977.

[15]   R. O. Lindseth, Seismic Surveying for Petroleum - the First 50 Years, *CODATA Bulletin*, Pergamon Press, p. 37, May, 1986.

Chapter 8

Architecture of a Database

In the previous chapter we examined structures suitable for storing technical data in a computer system to permit rapid access and efficient updates. This provided a description of the database at its lowest levels, closest to the computer hardware. In this chapter we look at the complete architecture of a database, from these lowest levels to the user's view, from the database software to the metadata used to assist database administration. This overview of a database system will help us to understand the specific data models used in actual database systems in the next chapter, Chapter 9, and how to go about planning and designing a technical database in the following two chapters, i.e. in Chapters 10 and 11.

8.1 SCHEMAS OR VIEWS

In Figure 8.1 we illustrate a generalized architecture for a database system. There have been many similar diagrams illustrating various database systems in the literature and there have been standard architectures published which arose from the work of the CODASYL committee in the 1960s and early 1970s [1] and from the later work of the ANSI/SPARC committee [2]. One essential feature of these architectures and of the illustration in Figure 8.1, is that they show either three or four views of the database (referred to as schemas). The four views in Figure 8.1 are

> (i) the *physical* schema which views the database as a collection of bytes and bits stored in blocks on disks (and possibly tapes),
> (ii) the *internal* schema which views the data as a set of logical files,
> (iii) the *conceptual* schema which takes an overall view of the database and its relationships, and
> (iv) the *external* schema representing the users' views of the data.

When only three views are represented in diagrams, the internal schema and conceptual schema are often combined together into one conceptual schema. This represents the simpler structure of many smaller database systems, particularly important for science and engineering. Sometimes the physical schema is omitted

"

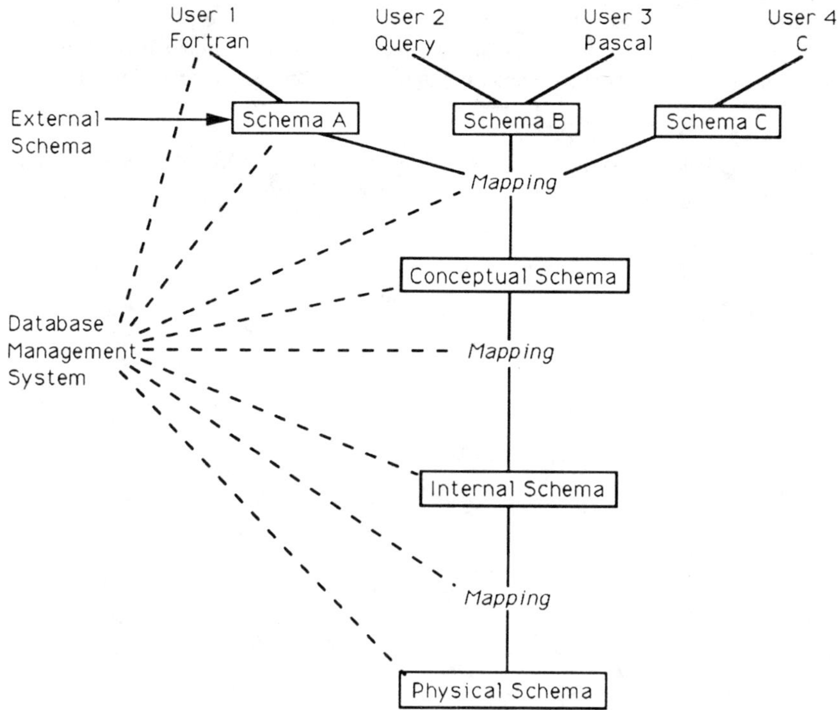

Figure 8.1 General architecture of a database.

from the architecture as the normal user does not need to have access to the data at this level.

To understand these we need to look at the different levels in greater depth.

8.1.1 The physical schema

The physical schema views the data in terms of tapes, disks, cylinders, tracks, blocks, bytes and sometimes bits. For example, the Atomic-Mass file discussed in the last chapter and described in Table 7.1 may be stored as variable length records. In Table 8.1 we illustrate how this data may be seen at the physical level. We have assumed that the first two records are stored together at the end of our block and that records 3, 4, etc are stored at a different location on the disk. In practice, of course, this is unlikely for such a short file. Records of a file are normally stored together in the same region of the disk and preferably within the one cylinder to make consecutive processing of the file efficient. However, long files are sometimes scattered across a disk, because the disk was fairly full when the file was first input or because frequent updates were made when the disk was full. As described in Section 7.3.3 when an insertion is attempted to a block which is already full, some

Table 8.1 Example of the Physical Schema or view of the Atomic-Mass file of variable length records corresponding to Table 7.1. The block size is assumed to be 1024 bytes, so block 83 could not hold more than two records. Therefore the remainder were stored elsewhere, in block 49. Note that only the Byte and Length fields need be recorded when the file is serial.

Record	Field	Disk Transport	Cylinder	Track	Block	Byte	Length
1	1	4	27	191	83	982	8
1	2					990	1
1	3					991	1
1	4					992	8
2	1					1000	6
2	2					1006	2
2	3					1008	1
2	4					1009	8
3	1	4	27	62	49	0	9
3	2					9	2
3	3					11	1
3	4					12	8
4	1					20	6
4	2					26	1
4	3					27	1
4	4					28	8
etc.							

of the records in the block must be moved to a new block, possibly on the other side of the disk. It is in this way that the file comes to reside on several parts of the disk.

The physical storage of fixed length records is simpler than for variable length fields. Let us assume that the file is stored in two separate locations on the disk, each covering several consecutive blocks. Then the physical schema needs only to record the positions of the first record and number of records in each area as in Table 8.2.

Table 8.2 Example of a physical schema or view of the Atomic-Mass file in Table 8.1 stored as fixed length records in two locations.

Format: Record Length 22 bytes
Field 1 Format A8; (A is Alphanumeric character)
Field 2 Format A2;
Field 3 Format I2; (I is Integer)

	Disk Transport	Cylinder	Track	Block	Byte	Length (records)
Location 1	4	27	191	83	982	2
Location 2	4	27	62	49	0	92

The system can then easily compute the position of each record. Normally records would not be split between two blocks, with part of one record at the end of one block and the rest at the beginning of thc next. Each block holds a number of whole records even though this leaves some unused space at the end of some blocks.

A file can also be stored in a chain structure where each area ends with a pointer to the next area as in Figure 8.2.

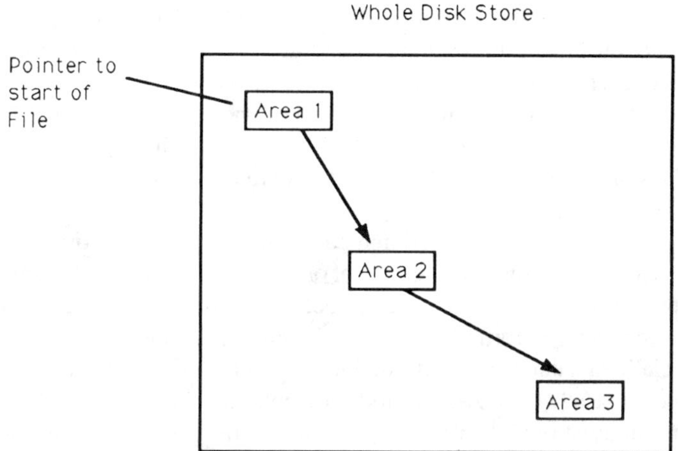

Figure 8.2 Storage of a file as a chain of three areas linked together with pointers. Each area may consist of several blocks and can hold fixed or variable length records (the latter as in Table 8.1 or separated by delimiters see Table 7.3).

8.1.2 The internal schema

The internal schema views the data in terms of logical files and records and in terms of the access mechanisms used to find records. For example, the Atomic-Mass file would be viewed as a set of records numbered 1 to 98. The fact that the file is actually stored in blocks and that it is stored in two positions on the disk is not visible at this level. The database software (i.e. the database management system) maps this internal schema onto the physical schema.

The internal schema is also concerned with access mechanisms. For example, if searches in the Atomic-Mass file are normally based on the atomic symbol then a method of finding the record for any symbol, say Pb, would be needed; for example this might be a B-tree (see Section 7.3.4). Then the internal schema would include a B-tree based on a sequence set in alphabetic order of the atomic symbol. The pointers to records would point to logical records, i.e. to record 42 or to record 67. These would be mapped at the physical level to physical addresses such as to cylinder 27, track 2, block 52, byte 81 on a disk pack.

8.1.3 *The conceptual schema*

This is a global view of all of the data in the database. In a business database this is used to provide a data model of an organization or part of an organization. In science and engineering it is used to provide a model for part of the physical world covered by the data (e.g. for the mechanical properties of alloys or for the numerical values of data in atomic physics).

Usually the conceptual schema is described by the user in terms of systems diagrams (discussed in Section 10.1.2) or in entity–relationship diagrams (discussed in Section 11.2). But few database systems allow any formal representation of such an overview of the database system, except, perhaps, as part of the data dictionary (see Section 8.3). In most systems the conceptual schema as stored by the database system is no more than a list of files, records, their fields, relationships between files, integrity constraints (so that a character does not occur in an integer field) and access constraints.

Although not normally included in a conceptual schema, it would be useful to include codes to enable the computation of one data item from another, for example to compute a Maxwellian averaged rate coefficient from a cross section or to find the resistance of a wire from data on the resistivity of the metal and its dimensions. A few specialized database systems do include such code, for example the ACTIS Tribology database [3], the Atomic and Molecular database in Belfast [4], and the FACT Thermodynamics database [5]. However, no general system (apart from experimental object-oriented systems, see Section 9.6.2) includes any facility to allow the addition of code. As explained in Section 7.6, you will have to add such enhancements yourself and if the code is simple this is not difficult.

8.1.4 *The user view*

The user view is sometimes called the external schema or subschema and it represents the user's view of the conceptual schema. This is normally a subset, which is sufficient to provide a model for part of an organization, or in a scientific or engineering environment for part of the physical world, a part which will enable the user to obtain answers to queries or to get some needed information from the database. One conceptual schema can support many user views so that a database can meet the needs of several classes of users.

To illustrate further what we mean by a user's view, we take an example concerning properties of materials. A conceptual schema may consist of a list of properties in tables, with the following fields:

> Material, Temperature, Electrical Resistivity, Magnetic
> Permittivity, Thermal Conductivity, Permittivity, Heat Capacity.

There may be many materials, and for each material there may be measurements at several temperatures.

Let us assume that you are an electrical engineer and that you are interested only in the electrical and magnetic properties of metal alloys at normal room temperatures. So your view of the database will consist of files of records with fields:

Metal Alloy, Temperature, Electrical Resistivity, Magnetic Permittivity.

Then any query to the database using your view or any programs accessing the database using your view can be written as if the files are in this simpler form. It would also be impossible for you to access the thermal conductivity or heat capacity (which you do not need) or to obtain resistivity data on other materials not amongst the subsection of materials, metal alloys, selected by your view. Therefore, if you were adding or modifying data as well as searching for it, it would be impossible for you, using this view, to erroneously change in any way the data on the other parts of the database not in your view, i.e. you could not change, even accidentally, any of the heat capacity data or any data (say) on composites. So this facility makes the other data in the database more secure.

However, few users need to write to the database and therefore few have a user's view which enables them to change it. To most users a database is a read-only system for accessing data. For these users, the main purpose of a view is to simplify the database, to make it easier to search and display, and to make it easier to write programs to access the data which are needed.

To search for data in a database, a view must be defined either beforehand or by the user. Some databases support searches only through predefined views; others allow the definition of views at search time; i.e. *ad hoc* views. In non-technical databases that contain business information, searching by predefined views permits considerable control. For example, a clerk receiving time cards is not allowed to access payroll information.

For many technical databases, scientists and engineers who use databases need and often demand access to the entire contents of a database, then separate predefined views are unnecessary.

8.1.5 *User interface*

Once a user view has been chosen, access is either through a special query language provided as part of the database system or through special code for searching in the database built into a user program. The special query language or the special code embedded in a program will be written in a language made available with the DBMS; the most common language is SQL (see Section 8.2.2).

If code is to be embedded into a user program then an interface must be provided by the DBMS for the language in which the program is written. The best database systems provide interfaces for several languages (e.g. FORTRAN, C, PASCAL) but many only provide an interface for one language (e.g. COBOL). Some provide no direct interface for external user's programs, but these will almost always provide within the database software a user language which includes simple arithmetic statements, loops, conditions and other structures needed to build a simple program (e.g. such facilities are provided in the most common PC database system dBASE). However, since these programming constructs are not part of any recognized language they are less useful than an interface to a FORTRAN or C or other language.

When no interface to the programming language which you want to use is provided (say FORTRAN), you can normally make the database system write an external file containing the data you need, and you can then put READ statements in your FORTRAN program to read the file. Usually this is not difficult. But since it involves two separate actions:

   (i) to interrogate the database to produce the file and
   (ii) to run the FORTRAN program

it is slow and clumsy compared with a system which allows a direct interface.

8.1.6 Smaller database systems

Not all database systems have all four levels of schemas described in Sections 8.1.1 to 8.1.4. In some the conceptual schema is no more than the total collection of files in the internal schema or it may be the totality of all of the users' views mapped onto all of the internal schemas. Sometimes, therefore, there are only three levels; the user view, which maps directly onto the internal schema, which then maps onto the physical schema. In many small systems, like dBASE III, there are no separate user views, only an internal or single user schema which is a set of files created by users and then the physical schema, the physical representation of these files.

8.2 DATABASE MANAGEMENT SYSTEM

The database management system is the software controlling the whole database. It is responsible for all of the operations and mappings (i.e. transformations) illustrated in Figure 8.1 including

   (i) the interface between users' programs and the database system,
   (ii) mappings from the users' views to the conceptual schema,
   (iii) mappings from this to the internal schema and
   (iv) mappings of the logical files of the internal schema onto the actual disks and tapes, i.e. onto the physical schema.

Since much of this is often handled by the operating system we discuss the relationship with the operating system next.

8.2.1 Operating system

Normally the database management system software runs as a large program on top

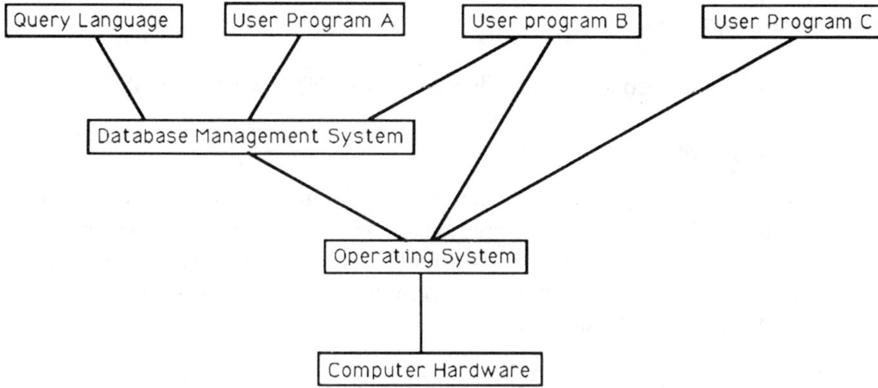

Figure 8.3 Usually the DBMS runs on top of the normal operating system on a computer and it may share the time available with other user programs (e.g. Program C).

of the normal operating system of a computer (see Figure 8.3). For example, the microcomputer database system dBASE may run on top of the operating system PC DOS on an IBM PC, or the database system INGRES may run on top of the UNIX operating system on a VAX. This enables the database system to share the facilities already available in the operating system and it makes it easier for user programs, running under the operating system, to gain access to the database.

In Figure 8.3, the user program C is accessing the computer system through the operating system at the same time as the DBMS; if both were running simultaneously they would be timeshared. On the other side of the diagrams there is a query language (which is often part of the DBMS software, see Section 8.2.2) and a user program A which are only using the facilities available in the DBMS. For example program A may be written entirely in whatever DBMS language is made available to users. However, program B is running under the operating system and at the same time can access the database through the DBMS to obtain data. So program B may be in FORTRAN and may include procedures or special CALL instructions to retrieve data. As we shall see later these special instruction are written in a special language, the data manipulation language.

Looking again at Figure 8.1 we can see that although the DBMS is responsible for all operations on the database and for all of the features in Figure 8.1, some of these may be carried out with the help of the operating system; an example may be the interface with users' programs at the top of the figure. Similarly at the bottom of the diagram, the mapping of the internal schema onto the physical schema is a normal function of an operating system; so this operation can be transferred from the DBMS to the operating system using a procedure or macro call to the operating system. Of course, this call is made automatically by the DBMS and is not seen by the user.

8.2.2 Data languages

We have already mentioned the data languages available with a DBMS.

1. Query language

The most important language, and often the only language many users see, is the query language. This is usually a simple system where you are asked to respond to a series of multiple choice questions. Normally it is self-instructive; so if you were unsure how to proceed at any point you only have to type HELP or ? to obtain assistance. Simple queries such as

> Atomic weight of lead from the Atomic Physics database, or
> Resistivity of copper from a Properties of Metal Alloys database

can be posed quickly by someone who is not an expert. A query such as

> Alloys with resistivity $10^7\,\Omega\,cm$ and melting point $3\,000\,°F$

would require slightly more expertise. (An example of a search using a query language in an atomic database is described in detail in the next chapter, see Figure 9.11(*b*)).

2. Data manipulation language (DML)

The data manipulation language consists of calls or instructions which can be built into a user program to obtain data from the database. These would enable requests for data to be expressed in the form of commands (rather than responses to multiple choice questions as in the query language). For example, in the most commonly used language in databases, SQL, (now an international standard) we would find the atomic weight of lead using the commands in Figure 8.4. These would be made part of the user (FORTRAN) program and would be compiled with the rest of the programs to include the atomic weight of lead within the program.

　　We discuss this language again and other languages suitable in the various data models in Chapter 9.

```
SELECT    Atom-Wt
FROM      Atomic-Mass
WHERE     Atom-Name = 'Lead';
```

Figure 8.4 Commands in SQL to search for the atomic weight of Lead.

3. Data definition language (DDL)

The least commonly used DBMS language is the data definition or data description language because it is generally used only once per file. Before data are entered into any file, the DBMS must first know:

   (i)   the number of fields there will be in each record,
  (ii)   the length and type of each field,
 (iii)   the primary key which can be used to identify each record,
 (iv)   alternate keys (if any),
  (v)   any additional fields or combinations of fields which should be indexed and
 (vi)   any further constraints on the data.

A simple example of a constraint might be that the symbol field in the Atomic-Mass file always consists of a capital letter, on its own or followed by a lower case letter. Further constraints might be that the atomic weight is always greater than the atomic number, and that the atomic number ≤ 98.

All of this information is read into the DBMS using one form of the DDL suitable for defining files in the conceptual and internal schema. For example, in the DDL in SQL we might describe the Atomic-Mass file, apart from the above constraints, using the instructions in Figure 8.5(a). The equivalent instructions in the microcomputer database system dBASE are given in Figure 8.5(b). Note the similarity, and how straightforward this operation can be in modern languages.

A second version of the DDL is used to create a user schema or view of the database. Let us assume, for example, that you wished to define a view consisting only of the symbol and atomic weight for the light atoms (atomic number ≤ 20). In SQL this would use the instructions in Figure 8.6(a). There is no provision for separate views in dBASE III or in most other microcomputer and other small database systems. However, in Figure 8.6(b) we give a second example for the language QUEL, a language which is part of the INGRES database system, used

```
User:        CREATE TABLE        Atomic-Mass
                                 (Atom CHAR (15),
                                  Symbol CHAR (2),
                                  Atom-No SMALL INT,
                                  Atom-Wt FLOAT);
```

Figure 8.5(a) Example of the data description language (DDL) in the language SQL needed to set up the Atomic-Mass file (see Table 7.1).

```
User:        CREATE ATOMMASS

dBASE:       Field Name          Type            Width    Dec
             Atom                Character         15
             Symbol              Character          2
             Atom_No             Numeric            6
             Atom_Wt             Numeric           19      3
```

Figure 8.5(b) Instructions in dBASE for the example in (a). Note that the dBASE response has been shortened for the sake of brevity. Dec is set to 3 for Atom_Wt to indicate three decimal points.

```
CREATE VIEW        Light-Atoms (Symbol, Atom-Wt)
SELECT             Symbol, Atom-WT
FROM               Atom-Mass
WHERE              Atom-No < 20;
```

Figure 8.6(*a*) Creation of a view of the file in Figures 8.5(*a*) and (*b*) using the data definition language in the language SQL.

```
RANGE OF Am is Atom-Mass

DEFINE VIEW Light-Atoms (Am.Symbol, Am.Atom-WT)

WHERE Am.Atom-No < 20
```

Figure 8.6(*b*) The example in (*a*) using the QUEL version of a DDL. (QUEL is a language used in the INGRES database system.)

widely on VAX computers. These examples illustrate that just as 'creation' in the conceptual DDL is relatively straightforward; so is the creation of views in these modern languages.

8.2.3 *Security, privacy, statistics*

The database management system has a number of other functions besides providing mappings between schemas, interfaces with user programs and data languages. It must also keep data secure by maintaining back-up copies of user files which it can load back onto the disk in case of machine failure. The ability to restart the system in the middle of a search or update may also be provided. In large systems where concurrent access to the database is permitted it will also ensure that one user cannot obtain inconsistent data because a second user is updating the same files or indices being accessed.

It will also guard against deadlock. In the simplest case deadlock occurs when one program is denying access to file A by other users while it waits for a data item from file B, but a second program is denying access to file B while it waits for a data item from file A. Of course, both would wait indefinitely and hold up access to files A and B to all users unless the DBMS (or operating system) was able to detect the deadlock and unscramble it, allowing first one program then the other to complete its computations.

The DBMS may also maintain the privacy of the data, using passwords or security codes to ensure that each user can access only those files to which the user is permitted access. Usually access is checked on whole files, but it may be checked on specific records or fields. For example, a museum database may be available for searching by the general public or by scholars; but the public schema may exclude, for example, many minor specimens duplicated in the main collection. It may also exclude the price field from the public schema so that unauthorized persons cannot find out which specimens are valuable and which are not.

File Name: *Atomic-Mass*;
Maximum Number of Records: 98;
Number of Fields: 4;

Field 1:   *Atom*; Char (15), Alternate Key. This is the name of an atom.

Field 2:   *Symbol*; Char (2), Key. This is a unique symbol representing the atom. It consists of a capital letter, sometimes followed by one lower case letter.

Field 3:   *Atom-No*; Integer (2), Alternate Key. This is the atomic number, the number of protons in the nucleus of an atom. Each atom has a unique value between 1 and 98.

Field 4:   *Atom-Wt*; Real. This is the atomic weight of an atom in atomic units. It is always a number greater than Atom-No.

Figure 8.7 Example of Metadata in the data dictionary, corresponding to the Atomic-Mass file in Table 7.1 and Figures 8.5(*a*) and (*b*).

The DBMS may also normally keep statistics not only on time and resource expended by users, for billing purposes, but also statistics on which files, records, fields and relationships are used most often. This enables the database administration (see Section 8.4) to tune the system to obtain rapid responses to the most common and important queries.

Lastly the DBMS provides access to the data dictionary, discussed in the next section.

8.3   DATA DICTIONARY

A data dictionary provides information on the data in the database. It does not describe the values of the data items in the records, but rather:

   (i) the names of the files,
  (ii) the names of the fields in each file,
 (iii) their formats, what they are and
 (iv) constraints.

An example of a record in a data dictionary is given in Figure 8.7.

The data dictionary is itself a database, smaller than the database it is describing, but it can be searched or updated in much the same way as any other database. Sometimes the data dictionary is included within the database it is describing; it then includes within itself a description of itself.

The kind of data about data held in the data dictionary are called metadata. They might also include:

    (v) the conceptual schema,
    (vi) information on the mappings between schemas,
    (vii) who uses which schemas,
  (viii) who can use which files,
    (ix) which programs use which part of the database (in case of changes) and
    (x) statistical information on usage and performance.

The metadata should be available for interrogation by any user, but probably the person to use them most often would be the person in charge of the database, the database administrator (see Section 8.5 below).

In brief, the data dictionary documents the database.

8.4   DATA THESAURUS

Whereas the data dictionary is concerned with metadata, the data thesaurus is concerned with the actual data in the database and is meant to be an aid to the user during a search. Very often a user who is not familiar with a particular database will be unaware of the terminology used in the database. The thesaurus helps the user (as illustrated earlier in Section 2.8); it can suggest possible alternate terms to the ones first selected by a user.

An example is illustrated in Figure 8.8 which concerns the CODATA referral database, an international database containing records on data centres providing data services in science and technology, covering a wide range of scientific and engineering disciplines. In a recent search a user wished to find databases on geology and on the first search missed the record in Figure 8.8 because it used the term geoscience, not geology. If the user had asked for help from the thesaurus, an entry as in Figure 8.9 would have been displayed and the user would then have been prompted to search for geoscience and any other of the words in the entry which might help find the records required. Thesaurus facilities are useful in large databases but they have often been implemented only in text databases or information retrieval systems (see Section 9.5).

8.5   DATABASE ADMINISTRATION

We have already mentioned that the database administrator is the person responsible for the maintenance and operation of the database. In a large commercial system which includes a large number of centrally held but interrelated files from many sources, the database administrator may have to have a large staff, as large as the systems support staff, to maintain and operate the database. In most technical database systems, however, the database administrator will be the scientist or engineer in charge.

```
Record 644
    SERIAL: (a) aa 812.
    COUNTRY: International.
    MAIN CATEGORY: Science and Technology (General).
    TITLE: World Data Center for Solid-Earth Geophysics.
    ADDRESS: National Geophysical and Solar-Terrestrial Data
        Center, NOAA, EDIS, Boulder, Colorado 80303 (USA).
    TELEPHONE: 303) 497-6215, ext 6474.
    TELEX: SOLTERWARN BDR 45897.
    INSTITUTION TYPE: International Organization.
    DIRECTOR: Lander, James F.
    SPONSORS: ICSU Panel of World Data Centres; US National
        Academy of Sciences
    COVERAGE: Data collection and services in the fields of
        seismology, tsunamis, recent crustal movements, heat
        flow, volcanology, magnetic measurements, gravity,
        earth tides, paleomagnetism and archeomagnetism.
        Exchanges are detailed in the 'Guide to International
        Data Exchange through the World Data Centres' by the
        ICSU Panel on World Data Centres.
    KEYWORDS: Geophysics; earth sciences; seismology;
        volcanology.
    OUTPUT: Publication of printed compilations; publication
        of data bulletins; Magnetic tapes containing data;
        Microfilms containing data.
    SERVICES: Provision of specific data upon request; Referral
        to institutions published data sources.
    AVAILABILITY: Open to all users.
    LANGUAGE: English.
```

Figure 8.8 Example of a record in the CODATA referral database not found by a search for the word Geology. A thesaurus entry, as in Figure 8.9 would prompt the user to search for Geophysics or Seismology which would have found the record.

```
GEOLOGY, geologies, geologist, geologize, geologise;

    Related Words:    geoscience,      geochemistry,
                      geocentre,       geodynamics,
                      geography,       geomagnetism,
                      geophysics,      geomorphic,
                      earth,           rock,
                      soil,            water,
                      environment,     seismology,
                      earthquake,      volcanology,
                      tsunamis,
```

Figure 8.9 Possible entry for the word Geology in a thesaurus for the CODATA referral database on data sources in science and technology.

The main responsibilities of the database administrator (DBA) concern the operation, maintenance, security, privacy and efficiency of the whole database system. All access to the database must be through the DBA, who must ensure, particularly for users allowed to input or edit files, that no user can change any part of the system needed by other users. The DBA establishes and maintains the data dictionary and consults the dictionary frequently to help his work, particularly when changes were being made to the database. Finally the DBA trains users and helps with planning and design when new additions are to be added to the database.

One of the important roles of the database administrator is the maintenance of the efficiency in a database system. An example follows in the next section.

8.6   EXAMPLE

We illustrate some of the points made in this chapter with an example. Consider an engineering database and included in the conceptual schema are the measurements in Table 8.3 of the safe axial loads for standard pipe columns for different pipe diameters and column lengths.

At this conceptual level the data would be represented by an array of values as in the table. But because the database system cannot store arrays the database administrator may have decided to store this table as a long file as shown in Figure 8.10 with the file sorted first on pipe diameter. Two user views of the data are illustrated in Figure 8.11, one in which the pipe diameter was fixed and the engineer searched and worked with the data as if it was a file at different column heights (as

Table 8.3 Safe axial loads for standard pipe columns in kips (stress in pipe, 75% of allowable stress).

Outside Diameter (in)	Length of column (ft)										
	6	7	8	9	10	11	12	14	16	18	20
3.500	25.3	24.2	22.8	21.4	19.7	17.9	15.9	12.4	9.2	6.9	
4.000	31.5	30.4	29.0	28.0	26.4	24.7	22.8	20.9	17.0	11.7	9.2
4.500	37.8	36.8	35.7	34.5	33.0	31.5	30.0	26.1	21.8	17.6	14.3
5.563	52.5	51.8	50.8	49.7	48.7	47.2	45.8	42.3	38.5	34.2	29.4
6.625	69.0	68.3	67.5	66.3	65.2	64.2	62.7	60.0	56.8	52.5	47.8
8.625	91.0	90.5	90.0	89.0	88.5	87.5	86.3	84.0	81.6	78.7	75.0
10.750	115	115	114	114	113	112	112	110	108	106	103

Diameter (in)	Length (ft)	Load (kips)
3.5	6	25.3
3.5	7	24.2
3.5	8	22.8
3.5	9	21.4
3.5	10	19.7
3.5	11	17.9
3.5	12	15.9
3.5	14	12.4
3.5	16	9.2
3.5	18	6.9
4.0	6	31.5
4.0	7	30.4
4.0	8	29.0
—	—	—
—	—	—
—	—	—
10.75	16	108
10.75	18	106
10.75	20	103

Figure 8.10 Possible internal schema (i.e. logical file structure) used to represent the data in the conceptual schema (overall view of the data) in Table 8.3.

View 1 **Load for Fixed Diameter d(in)**

As

```
SELECT Length, Load

FROM File in Figure 3.10

WHERE Diameter = d;
```

View 2 **Load for fixed Length L(ft)**

As

```
SELECT Diameter, Load

FROM File in Figure 3.10

WHERE Length = L;
```

Figure 8.11 Two possible user views of the data in Table 8.3 and Figure 8.10. The first is convenient when most calculations are for a fixed diameter; the second more convenient for a fixed length. View 1 is closest to the internal schema in Figure 8.10; so it is faster to execute than view 2.

in the internal schema) and the second being the reverse in which the engineer sees the data varying with diameter, but the column height remaining constant. The first view will give fast responses for each access to the database because the internal schema is similar to the user view of the data: all values for a constant diameter are stored beside one and other. But in the second view, which assumes the column height is fixed, every time a user accesses the virtual file of safe axial load versus diameter, and changes the diameter, the system would have to make another search through the whole file for the readings with the new diameter and then find the particular column height. This would be relatively slow compared with the first view.

Assume that after some time the data administrator notices that the statistics in the data dictionary show that he was mistaken when he set up the database and that more engineers are using the second view than the first view. Then to speed up the majority of searches and to economise on the use of resources the database administrator would change the internal schema to a file, sorted first on the column height as in Figure 8.12. A corresponding change to the physical schema would also be needed.

Of course with such a short file this rearrangement would make little difference in practice. In a long file covering a large number of blocks on disk, the difference in search times between the two views might be an order of magnitude.

This internal schema is an example of a many-to-many relation (many diameters each with many heights and vice versa) and we have represented this in the form of

Length (ft)	Diameter (in)	Load (kips)
6	3.5	25.3
6	4.0	31.5
6	4.5	37.8
6	5.563	52.5
6	6.625	69.0
6	8.625	91.0
6	10.75	115.0
7	3.5	24.2
7	4.0	30.4
—	—	—
—	—	—
—	—	—
20	8.625	75.0
20	10.75	103.0

Figure 8.12 Rearranged internal schema chosen to suit the majority of engineers who are found to use View 2 rather than View 1 in Figure 8.11. This new schema permits faster processing with View 2 at the expense of slower processing with View 1.

a table representing a file of records in Figures 8.10 and 8.12. But this structure is only appropriate in two types of database (called models), the Relational model and the Inverted List model. In either of the other two models in wide use, the Hierarchical model or Network model, the file would look quite different. We look at these models and a few others in the next chapter.

REFERENCES

[1] CODASYL Systems Committee. Introduction to Feature Analysis of Generalized Data Base Management Systems, *BCS Comp. Bull.*, **15**, April, 1971.

[2] ANSI/X3/SPARC Study Group on Data Base Management Systems, Interim Report. *FDT, ACM SIGMOD Bulletin*, **7**, No 2, 1975.

[3] S. Jahenmir, S. M. Shu and R. G. Munro, ACTIS: Towards a Comprehensive Technology Data Base, in *Computerization and Networking of Materials Data Bases*, Eds. J. S. Glazman and J. R. Rumble, Jr., Amer. Soc. Testing and Materials, Philadelphia, 340–8, 1989.

[4] J. G. Hughes and F. J. Smith, AMDS: A Database System for Atomic and Molecular *Physics, Comp. Phys. Comm.*, **32**, 317331, 1984.

[5] A. S. Pelton, Numerical Databases and Thermodynamic computations in Metallurgy and Materials Science, *Abstracts of the tenth International CODATA Bulletin*, No 61, p.67, May, 1986.

Chapter 9

Data Models

The data model for a database system determines the overall structure of the database management system. In a business environment this database model acts as a model for an enterprise or for a whole organization. In science and engineering it models the technical environment covered by the database. But for a technical database the term 'model' is not quite as appropriate as it is in the business environment. The idea of the model is that rather than use all of the file structures mentioned in the last chapter, with one chosen appropriately for each file, instead one basic structure is used for all data. This simplifies the task of writing a general database system for all data.

It follows that the term 'data model', used in every book on database systems is misleading. A better term would be 'data model facility', that is a facility or methodology to enable a user to produce a data model of an enterprise or of an area of science or technology. It is a facility to enable us to write a database system quickly. For this reason the ISO committee dealing with databases is recommending that the term 'data model' be replaced by 'data model facility'[24]. However, in keeping with all of the current literature on the subject, we will continue to use the term data model in this book.

There are three principal models:

   (i) Relational, where the data are represented by tables,
  (ii) Hierarchical, where the data are represented by trees, and
 (iii) Network, where the data are represented by many 2-level trees.

In (ii) and (iii) the data are structured mainly with pointers. We discuss these in more detail later in Sections 9.2 and 9.3 followed by the less common inverted list model in Section 9.4. Text or document databases are concerned with a different form of data than the other four; the corresponding text model for data is discussed towards the end of the chapter, Section 9.5. A further important model, the subject of much research, the Object-oriented model, is discussed last. However, because of the primary and growing importance of the relational model, particularly for micro-computers or mini-computers, we discuss it first and in much greater detail than the others.

9.1 RELATIONAL MODEL

9.1.1 Definitions

The mathematical description of the Relational model for a database [1] need not concern us here. For the scientist or engineer, a **relational database** is one where all of the data is represented in tabular form. Each table is called a **relation** and a collection of relations in one subject area is called a relational database.

The great advantage in the tabular structure, from the point of view of the user, is that it is simple and easy to understand. A relation is also very similar to the more commonly known structure, a file. Each row in the table is a record of the file and each column corresponds to a field of the record. In the terminology of relational databases each row of a table is a **tuple** and the heading of each column, corresponding to the field, is called an **attribute**. A data item at the intersection of any row and column is called a **component** of the corresponding tuple (or record). For example, looking back at the *Atomic-Mass* file in Chapter 7 (Table 7.1), each record of that file becomes a tuple, the attributes are '*Atom*', '*Symbol*', '*Atom-No*' and '*Atom-Wt*', the same as the fields. The first tuple (record) is made up of the components:-

Hydrogen, H, 1 and 1.008.

Before considering a second example let us define a few more terms for completeness. A **domain** of a particular attribute is the range of values which can be taken by the appropriate components. For example, the domain of the third attribute *Atom-No* in our Atomic-Mass relation is the set of integers between 1 and 98 (assuming we stop at Californium) and, the domain of the second attribute, *Symbol*, is the set of atomic symbols for the atoms: H, He, Li,..., Cf.

It is convenient to describe the relation in this example

Atomic-Mass (Atom, Symbol, Atom-No, Atom-Wt)

which is an example of the general form:

$$R(A_1, A_2 .... A_m)$$

where R is the name of the relation and A_k is the kth attribute. This is called a relational **scheme** and it is the usual way in which we describe a relation. A collection of these schemes for all the relations in a database is called the database scheme; it describes the whole database.

The number of attributes in a relational scheme (i.e. the number of columns in the relation) is called the **arity** or **degree** and the number of tuples (that is, the number of rows) is called the **cardinality**.

The collection of all these terms are listed in Table 9.1. However, although these new terms are used commonly by computer scientists, many relational database systems on the market do not make full use of these terms. It is common for better known terms, such as file, record and field, or alternatively table, row and column

Table 9.1 Terminology used for relational databases by computer scientists and the nearest equivalent used in data processing.

Relational Terminology	Nearest Data-Processing Terminology
Relation	File
Tuple	Record
Attribute	Field
Component	Data item
Schema	Format
Domain	Range, Type
Arity or degree	Number of fields
Cardinality	Number of records

to be used instead; this makes the systems easier to understand and probably increases sales. To make this book more easily understood by a reader who is not a computer scientist, we will normally use the simpler, more common terms.

Now let us look at a second slightly more difficult but often a more realistic example, the original data on ferrous alloys in Table 9.2(a). We first need to name the relation—we call it Hydrogen-Effect. Next we need to identify the attributes and at first sight these are fairly obvious:

Form
Condition
Hydrogen-Charging
Condition for Tensile Testing
F_{tu}
F_{ty}
e

But the footnote makes us think again. The information 'Hydrogen content 400–450 ppm' obviously must not be left out; but how should it be included? One possibility is that it be added to the field *Hydrogen Charging*, but this field already looks rather long. So it is better, because it is obviously another 'attribute' or property of the data, that it can be added as another field, i.e. we add a field *Hydrogen Content*. In addition, the information in the field *Hydrogen Charging* can be recorded as four separate attributes:

Fields		**Values**
Hydrogen Charging	Pressure:	0.025 A/cm^2
Hydrogen Charging	Temperature:	212 °F
Hydrogen Charging	Time	1 hour
Hydrogen Charging	Solution	1 N H$_2$SO$_4$ + sodium arsenite

We have assumed that the temperature of *Hydrogen Charging* is likely to change in other records. If it was constant for **all** records it could be taken out of the relation to save space.

If we looked at the first two records only, and ignored the third, the above set of fields would appear to be sufficient, but when we look at the third column we see that there is information present on the *Condition for Tensile Testings*, i.e. a delay of 16 hours at 392 °F. So we can add two new attributes or fields. The complete table viewed as a relation is shown in Table 9.2(*b*).

9.1.2 Matrices and relations

Many scientists or engineers are likely to think of a matrix when we describe a relation as a table. But a matrix and relation are not the same. A more exact description of a relation defines it as a *set* of tuples or records. As in any mathematical set, the order of the records in a relation is not important. So if we interchange two rows or reorder the rows, this does not effect the information stored within the relation. Of course, this would not be true of a matrix. Also if two rows of a relation are identical then they contain the same information; so one is redundant and should be removed. Indeed in many relational database systems, such redundant records are automatically removed. So if you store a matrix as a relation in a relational database, it might very well come out looking very different from the way in which you put it in.

If you do need to store a matrix, $B(I,J)$ for $1 \leq I \leq N$, $1 \leq J \leq M$, you need to store the number of the row (I) and column (J) of each value $B(I,J)$, i.e.

$$I, J, B(I,J)$$

An example is given in Tables 9.3(*a*) and (*b*).

Table 9.2(*a*) Effect of electrolytically charged hydrogen on tensile properties of foil (41).

Form	0.003" foil specimens		
Condition	1832 °F, anneal		
Hydrogen$^+$ Charging	0.025 Acm^2 at 212 °F, for 1 hour in 1 N H$_2$SO$_4$ + sodium arsenite		
Condition for tensile testing	No charging	H$_2$ charged	Charged† 392 °F, 16 hr
F_{tu} ksi	77	66	71
F_{ty} ksi	31	42	36
e, %	25	8	25

†Hydrogen content 400–450 ppm

Table 9.2(b) Some data as in Table 9.2(a) viewed as a relation. 'n.a.' is a special term for 'not applicable', possibly the null character. Although the data item for H_2-Charging is recorded either as 'no charging' or 'charging', it could be represented by the Boolean value False or True in the computer record.

Fields	Record 1	Record 2	Record 3
Form	0.003" Foil(41) Specimens	0.003" Foil(41) Specimens	0.003" Foil(41) Specimens
Condition	1832 °F Anneal	1832 °F Anneal	1832 °F Anneal
H_2-Charging	No charging	Charging	Charging
H_2-Content (ppm)	400–450	400–450	400–450
H_2-Charging Pressure (A/cms 2)	n.a.	0.025	0.025
H_2-Charging Temperature (°F)	n.a.	212	212
H_2-Charging Time (hrs)	n.a.	1	1
H_2-Charging Electrolyte	n.a.	1 N H_2SO + Sodium Arsenite	1 N H_2SO_4 + Sodium Arsenite
Tensile-Testing Delay (hrs)	0	0	16
Tensile-Testing Temperature (°F)	n.a.	n.a.	392
F_{tu} (ksi)	77	66	71
F_{ty} (ksi)	31	42	36
e (%)	25	8	25

Table 9.3(*a*) Uniformly distributed loads in pounds on rectangular beams 1" wide for fibre stress of 1 000 psi. For any other stress S multiply by $S/1000$.

Span	Depth of beam in inches						
ft	6	7	8	9	10	11	12
5	800	1090	1420	1800	2220	2690	3200
6	670	910	1180	1500	1850	2240	2670
7	570	780	1010	1290	1590	1920	2280
8	500	680	890	1120	1390	1680	2000
9	440	600	790	1000	1230	1490	1780
10	400	540	710	900	1110	1340	1600
11	360	490	650	820	1010	1220	1450
12	330	450	590	750	930	1120	1330
13	310	420	550	690	850	1030	1230
14	290	390	510	640	790	960	1140

Table 9.3(*b*) How the matrix data in Table 9.3(*a*) might be stored in a relational database. Rather than repeat the Stress and Beam Width (assuming it does not change) an addition can be added to the title of the relation in the data dictionary.

Stress (psi)	Beam Width (Inch)	Span (ft)	Beam Depth (inch)	Load (pounds)
1000	1	5	6	800
1000	1	5	7	1090
1000	1	5	8	1420
...				
1000	1	6	6	670
1000	1	6	7	910
...				
etc.				

9.1.3 Primary and alternate keys

Most database systems will also ask you to define the primary key of each file. This is the field or collection of fields which enables you to uniquely describe or identify each record. For example, in our Atomic-Mass relation the key could be the atomic name, the atomic symbol or the atomic number. If one of these was chosen, the others would be alternate keys. The atomic weight, a real number, is not likely to be chosen as a key although in this case it too would uniquely define the record.

Many implementations of relational database systems do not require the above definition of a key 'uniquely' describing a record. If you declare a key which is not unique, the system will still index the file on that key, usually by sorting. But it is better to identify a unique key.

Choosing a key for the Atomic-Mass relation was easy. For the Ferrous Alloy relation in Table 9.2(*b*) it is rather different. It requires all of the first 10 fields to describe the property data in the last three fields. So the 'key' in this case is

Form, Condition,....Tensile Testing, Temperature (°F)

and to identify any record in such a relation in a search, you would need to identify all 10 fields. Usually an engineer will not know all of these; he will specify as many as are known and a relational database system should be able to retrieve all those records which match the incomplete query.

Many relational (and other) database systems will automatically put all the records into some order, dependent on the initially chosen primary key. For example, the records might be put automatically into alphabetic order or numeric order (ref. Section 7.3.5), and then duplicates will be removed.

9.1.4 *Operations on a relational database*

There are a number of operations common to all relational databases (and indeed to all databases). These are:

1. Creating and changing the structure of the database (relations, fields, for-mats, schemas (views))
2. Inputting data and editing existing data.
3. Indexing
4. Searching and retrieval

The first three are discussed in separate paragraphs below and the last, searching and retrieval are described in the following section, Section 9.1.5.

1. Creation

The creation of a new database or of new relations within a database is carried out by the data description language (DDL) mentioned in Chapter 8, Section 8.2.2. This may be included in the software for data input, retrieval, etc or it may be a separate program.

Of course, before beginning to create relations in a database, you should first have planned what you want to do with the relations and designed a set of relations to meet your needs. These planning and design processes are discussed in the following two chapters.

Once you know what relations you wish to create, their fields and data types, the DDL is usually easy to use. You normally have to respond to a set of prompts or multiple choice questions. For example, you may be asked the name of the relation and then the number of attributes (fields), and next you will be prompted for the name of each field and the field type or format. Usually you have to choose a fixed length for each field and for field type you normally have a choice of integer, real number, alphanumeric character or Boolean. Sometimes a free text format allows you to enter comments. A simple example is given in Figure 9.1. Some special formats are also available in some database systems, formats such as 'date' or 'money' but these are of little particular interest to the scientist or engineer. Few systems allow the wider range of data types needed often by a scientist or engineer, types such as vectors or matrices; in addition you will normally not be asked for units. When units are not included they should be made part of the field names so that they appear in all print-outs and in menus used for data input.

For larger systems the DDL can be used to create all or part of the conceptual database (ref. Section 8.1.3.), including the relations in the internal schema and you, as a user, can also use the DDL to create your own particular view of the database. An example is given in Figure 9.2. The use of SQL and QUEL to create the user view in the same example was given in Figure 8.6.

2. Input and editing

Input and editing of records is generally straightforward. Once you have specified your relation by name (the active relation), you usually answer a multiple choice question on whether you wish to add data, edit data or delete data. For data input a menu is usually displayed showing the fields; and as you type in the data it is checked

```
Name of New Relation?                        Atomic-Mass

Number of Fields?                            4

Name of Field 1?                             Atom
Data Type? 1 Alph, 2 Int, 3 Real:           1
Number of Characters?                        20

Name of Field 2                              Symbol
Data Type? 1 Alph, 2 Int, 3 Real:           1
Number of Characters?                        2

Name of Field 3?                             Atom-No
Data Type? 1 Alph, 2 Int, 3 Real:           2
Number of Digits?                            2

Name of Field 4?                             Atom-Wt
Data Type? 1 Alph, 2 Int, 3 Real:           3
Primary Key?                                 Atom-No
Alternate Key?                               Symbol
Alternate Key?                               None
```

Figure 9.1 Example of the creation of our Atomic-Mass relation by response to a series of simple questions posed by the data definition language. The relation would be indexed on the atomic number, with a secondary index on the atomic symbol.

```
Name of user relation?                          Light-atoms

From which relations in internal schema?        Atomic-Mass

Select fields?                                   Symbol; Atom-Wt

Specify constraints?                             Atomic-Mass.Atom-No < 20
```

Figure 9.2 Example of the creation of a user schema by responses to a series of questions posed by the DDL.

to ensure that it has the correct format, for example, you should not be able to type an alphabetic character in an integer field.

An example of data input to the relation created in Figure 9.1 might be as illustrated in Figure 9.3.

For data editing, usually you first display the record you wish to edit on the screen, you make the change and immediately the new record is displayed, sometimes with a final check question such as: 'Is it correct?' which must be answered before the system changes the record in the file. A similar check is made before a record or relation is deleted. Because the mistaken deletion of data can be so traumatic, in many systems 'deleted' records are initially not physically deleted from the disk, but marked as no longer part of the file. Facilities to delete them physically or to restore them because of a mistake (if not physically deleted) are then also provided.

3. Indexing

When creating a relation you may be asked for the primary key; this is used for indexing. Examples are given in Figures 9.1 and 9.2. The database system will then automatically generate an index for the primary key. This is often a B-tree (see Section 7.3.4); but it may be a hashed index (see Section 7.3.7). This index then

```
Relation?                    Atomic-Mass

Record 1

Atom?                        Hydrogen
Symbol?                      H
Atom-No?                     1
Atom-Wt?                     1.008

Record 2

Atom?                        Helium
Symbol?                      He
Atom-No?                     2
Atom-Wt?                     4.003

etc.
```

Figure 9.3 Example of input of data to the Atomic-Mass database created in Figure 9.1

enables the records corresponding to particular primary keys to be found quickly with only a few disk accesses. In some systems, you have to specify that you need an index. If many searches are likely on the primary key, then an index should be created.

Sometimes when alternate search strategies are common then a second index is needed. For example, in the case of the Atomic-Mass relation in Figure 9.1 we have specified one alternate, the atomic symbol. Additional indexes add to the size of a database, but they also reduce access time. The optional number of indexes becomes a trade-off between access time and storage.

All of the operations described in the above sub-sections on creation, input, editing and indexing of data are simple and there is no universal language available with which to describe them all. For each system, you need to read the user manual to learn the exact syntax to employ. On modern systems, which are more user-friendly than a few years ago, they can all be learned in a few hours. However, the difficulty is not the syntax, but with decisions on which fields are in which relations. The example in Tables 9.2(*a*) and (*b*) illustrates that such decisions are not so easy and need a knowledge of the data, both present and in the future, and of the uses to which the data will be put. As stated earlier this is discussed in the following Chapters 10 and 11 on planning and design.

Searching and retrieval from a database is a little more difficult than use of the DDL or input of data, but for a relational database it is still straightforward. We discuss it next.

9.1.5 *Retrieval from a database*

As mentioned in Chapter 8, retrieval of data from the database is frequently possible by means of a special query language which is part of the database system and which can be used to find data in the database. In addition, in many systems a data manipulation language is available (ref. Section 8.2.2) which is a set of commands embedded within a user program which can access and manipulate the data. Unfortunately not many systems allow the commands to be embedded in a FORTRAN program, which is the most likely language a scientist or engineer will want to use. However, in almost all systems, the retrieved data can be stored in a file which can then be interrogated later by a separate user program.

1. *Selection*

Whether interrogation is via a query language or a user program the most common operation is the **selection** of a set of records S from a relation R which satisfies some condition F. In relational algebra [2] this is written

$$S = \sigma_F(R)$$

Note that S is also a relation. For example in the Atomic-Mass relation we may wish to select out the record with an atomic number of 92. By definition the order of records can be changed; so we do not know which record number holds the appropriate record and we therefore always search on the subject or data item in a record. This request can be written in the algebra as

$$S = \sigma_{Atom\text{-}No\,=\,92}(Atomic\text{-}Mass)$$

which would retrieve the Uranium record.

2. Projection

The second common operation is **projection**. This concerns the selection of columns or fields rather than records or rows of our table. It is written in the form

$$S = \Pi_{a.b.c}\,(R)$$

to project out the fields a, b, and c from the relation R. An example in our atomic physics relation is as follows:

$$S = \Pi_{Atom.Atom\text{-}Wt}\,(Atomic\text{-}Mass)$$

to project the two columns for the atomic name and atomic weight.

Commonly these two operations, selection and projection, are put together. For example, to project the above fields *Atom* and *Atom-Wt* for *Uranium* we write

$$S = \Pi_{Atom.Atom\text{-}Wt}\,(\sigma_{Atom\,=\,'Uranium'}\,(Atomic\text{-}Mass))$$

and S would be the single record relation

Atom	*Atom-Wt*
Uranium	238.07

Languages

In the various languages used in database systems, these two operations are written in many forms. However, all of these can be given a mathematical description in relational algebra and there are also two other equivalent descriptions in terms of relational calculus called tuple calculus and domain calculus, both similar to propositional calculus. A good description of these three mathematical representations of the database operators is given by Ullman [2] but this is outside the scope of this text. Instead we give examples in two languages SQL and QUEL which are based on the two relational calculus expressions. These languages are more easily understood by most readers than the mathematics. We also illustrate with the language of the most commonly used microcomputer relational database system, of dBASE, which has a structure not unlike SQL.

The first language we use for demonstration is SQL which has been developed by IBM, but is now widely in use on other systems and is effectively the international standard language for relational databases[3]. In SQL our Uranium example is written in the form:

```
SELECT     Atom, Atom-Wt
FROM       Atom-Mass
WHERE      Atom-No = 92;
```

The language is so clear in this case (and in most other cases) that it does not need further explanation.

In the second language QUEL, which is the language most often used with INGRES relational database systems by Relational Technology and which is available, for example, on VAX computers and used commonly by scientists and engineers, the Uranium example above would be written

```
RANGE OF x IS     Atom-Mass (Atom, Symbol, Atom-No, Atom-Wt)
RETRIEVE INTO     S(x.Atom, x.Atom-Wt) WHERE x.Atom-No = 92
```

The first line defines tuple variables x of *Atom-Mass* and the second line stores the appropriate fields *x.Atom*, *x.Atom-Wt* of the selected record or tuple.

Our third language is the one provided with the database system, dBASE, which is available on most microcomputer systems. Our Uranium example would be written

```
USE Atom-Mass
COPY TO T FOR Atom-No = 92 FIELDS Atom, Atom-Wt
```

The first line assigns file *Atom-Mass* to workfile *T* on which the following copy command operates. It needs no explanation.

3. Join

When you need to use two separate relations to obtain data it is most likely that you will use the Join operation. This usually joins two relations on one or more common field.

To illustrate this, let us return to an example in atomic physics. If you look at a table of the ionization energies of atoms you will normally see a table similar to Table 9.4 and we could create a database made of 2 relations – the *Atomic-Mass* relation in Table 7.1 and an Ionization relation made up of records in Table 9.4. Unfortunately this duplicates the information relating the *atom name, symbol* and *atomic number*. As explained in Chapter 8, one of the aims of a database system is to reduce such duplication to a minimum. So if we wished to store data on ionization

Table 9.4 Atomic ionization energies.

Atom	Symbol	Atomic No	Ionization Energy
Hydrogen	H	1	13.59
Helium	He	2	24.59
Lithium	Li	3	5.36
Beryllium	Be	4	9.28

energies we would store only one of the above 3 fields (probably atomic number) in a simpler relation:

Atom-Ionization (Atom-No, Ionization-Energy)

as in Table 9.5. This will keep redundancy to a minimum and remove the chance of inconsistency (e.g. Argon represented by the Symbol Ar in one relation, and by the symbol A in a second).

Table 9.5 The relation *Atom-Ionization*.

Atomic-No	Ionization Energy
1	13.59
2	4.58
3	5.36
4	9.28

The new relation might not correspond to a table which we would want to print. For example, in a printed table we would want to include the symbol and possibly also the atom name to make the table easier to comprehend. We produce such a printed table, not by duplication, but by joining the two relations and projecting out the fields we want to print. The join is on the common field, *Atom-No*: a record of the first relation is joined to a record of the second relation if the common field is the same, i. e., only records with the same atomic number are joined. So the joined relation has fields

Atom-No, Atom, Symbol, Atom-Wt, Ionization-Energy

To obtain the records in Table 9.4 we project out the four fields apart from *Atom-Wt*.

In relational algebra a join is represented by the symbol $|X|$; so in this algebra Table 9.4 becomes:

$$\Pi_{Atom.Symbol.Atom-No.Ionization-Energy}(Atomic\text{-}Mass \ |X| \ Atomic\text{-}Ionization)$$

Fortunately, the expression of this join in most languages is easier to understand than the mathematical expression. In SQL it is particularly easy to understand:

```
SELECT    Atom, Symbol, Atom-No, Ionization-Energy
FROM      Atomic-Mass, Atom-Ionization
WHERE     Atomic-Mass.Atom-No = Atom-Ionization.Atom-No
```

In the last line the expression

Atom-Mass.Atom-No

means the field *Atom-No* of the relation *Atomic-Mass*.

In the second sample language, QUEL, Table 9.4 is written:

```
RANGE OF u IS Atomic-Mass (Atom, Symbol, Atom-No, Atom-Wt)
RANGE OF v IS Atomic-Ionization (Atom-No, Ionization-
Energy)
RETRIEVE (u.Atom, u.Symbol, u.Atom-No, v.Ionization-Energy)
WHERE u.Atom-No = v.Atom-No
```

The first two lines define records u and v in relations *Atomic-Mass* and *Atomic-Ionization*; the logical condition after WHERE effects the join and the expression in brackets after RETRIEVE projects the columns we need and also prints the table with headings *Atom, Symbol, Atom-No, Ionization-Energy*.

In the microcomputer dBASE system, Table 9.4 is written:

```
SELECT          2
USE             Atomic-Mass ALIAS AM
SELECT          3
USE             Atomic-Ionization ALIAS AI
JOIN WITH       AM TO T
    FOR         AM —> Atom-No = AI —> Atom-No
    FIELDS      Atom, Symbol, Atom-No, Ionization-Energy
```

The first two lines assign file *Atomic-Mass* to workfile 2 and allow us to use the shorter alias *AM* instead of *Atomic-Mass* in the following commands. The next two lines assign file *Atomic-Ionization* to workfile 3. The last command joins this file (the active file in the previous line) with file *AM* (i.e. *Atomic-Mass*) and copies records to *T*. The second line performs the join: *AM —> Atom-No* means field *Atom-No* of *AM* and the third line specifies the fields to be copied to *T*. This command is not easy to follow.

Note that in all three languages we could add further conditions after the WHERE or FOR operators making the join a conditional join. For example, to print records only for the light atoms (atomic number <17) the QUEL expression after the operator WHERE would be:

```
WHERE u.Atom-No = v.Atom-No AND u.Atom-No <17
```

Joins, similar to the one we have just described, are the principal method available for linking together the data in two or more separate relations. In other modules, as we shall see, such linkages are achieved using pointers. In the relational model the only method of connecting together data items not in the one relation is through the use of logical operations on common fields in pairs of relations. We describe some other operations of this kind.

4. AND, OR, NOT

Since relations are defined as sets of records, the usual operations on a set, $\cap$(AND), $\cup$ (OR) and $\neg$ (NOT), can be used on two relations to create a new relation provided the two relations have the same fields (or have fields with overlapping domains, e.g. the fields may be light atoms in one relation and all atoms in a second).

Union. If X and Y are two relations with the same fields, then:

$$T = X \cup Y \quad \text{or} \quad X \text{ OR } Y$$

is the relational algebra expression for the relation T made up of records which belong to X or Y, or both.

In most languages this is implemented using an APPEND command, as in QUEL (in the following examples we assume that X and Y both have fields A and B):

```
RANGE OF        x IS X (A, B)
RANGE OF        y IS Y (A, B)
RETRIEVE INTO   T(x.A, x.B)
APPEND TO       T(y.A, y.B) WHERE y.A = x.A OR y.B = x.B
```

In dBASE this would also be written using an APPEND operator:

```
SELECT 2
USE S
SELECT 3
USE R
COPY TO T FIELDS A,B
USE T
APPEND FROM S FOR S → A = R → A OR S → B = R → B
```

This still includes duplicates; these can be removed using the SET UNIQUE ON command. In SQL this is clearer, if a little clumsy:

```
SELECT A, B
FROM X
UNION
SELECT A, B
FROM Y;
```

Intersection. The expression

$$T = X \cap Y \text{ or } X \text{ AND } Y$$

is the relation made up of records belonging to both X and Y. The QUEL expression for intersection is:

```
RANGE OF x IS X(A,B)
RANGE OF y IS Y(A,B)
RETRIEVE INTO T(x.A, x.B)
WHERE x.A = y.A AND x.B = y.B
```

In dBASE we use a Join operator:

```
JOIN WITH S TO T FOR A = S → A AND B = S → B
```

which comes after the same SELECT and USE statements as in the union example. In SQL the expression can take a similar form:

```
SELECT x.A, x.B
FROM X, Y
WHERE x.A = y.A and x.B = y.B
```

Deletion. The expression

$$T = X - Y$$

is the relation T of records belonging to R which do not belong to S. In QUEL this can be written, as in many languages, using a DELETE statement, i.e.:

```
RANGE OF x IS X(A,B)
RANGE OF y IS Y(A,B)
RETRIEVE INTO T(x.A, x.B)
RANGE OF t IS T(A, B)
DELETE t WHERE x.A = y.A AND x.B = y.B
```

The dBASE expression should be similar, i.e. after numbering the instructions:

```
1 SELECT 2
2 USE S
3 SELECT 3
4 USE R
5 COPY TO T FIELDS A, B
6 USE T
7 DELETE FOR A = S → A, B = S → B
```

In our version of dBASE III this only deletes the first record of R which also belongs to S, which shows, at least in this example, that dBASE III does not treat relation R as a **set** of records. (So dBASE III is not always relational). We have to write the following clumsy commands after the first five commands above:

```
 6 INDEX ON S → A, S → B TO R-INDEX
 7 SELECT 1
 8 USE T
 9 SET RELATION TO S → A, S → B INTO S
10 DELETE FOR R → A = S → A AND R → B = S → B
```

For the meaning of these commands the reader should refer to a dBASE manual.

In SQL the expression is not quite so easy to understand. It is:

```
SELECT A, B
FROM X
WHERE NOT EXISTS
      (SELECT A, B
      FROM Y
      WHERE X.A = Y.A AND X.B = Y.B);
```

To understand this expression, let us examine it step by step. In the first two lines we select a record of X, i.e. $X.A$, $X.B$ and we find whether it satisfies the condition after WHERE. This is TRUE only if the subquery after NOT EXISTS is empty, i.e. only if there does not exist a record $Y.A$, $Y.B$ of Y which is identical to the record $X.A$. $X.B$ of X. This is the same as the condition in the definition of $X - Y$. It follows that the above SQL expression is a relation equal to $X - Y$.

In this example we have introduced the nested subquery and a NOT EXISTS operator of SQL to illustrate the power of the SQL language. It should be mastered by a scientist or engineer, but we think that this is a difficult language for non-technical people. A full description is outside the scope of this book. A complete list of operations and a comprehensive list of the types of logical queries common in a business environment is given by Date [4] with examples in both SQL and QUEL. However, most of these are not frequently needed in science and engineering while many operations which are needed, such as interpolation in a table, are not available.

9.1.6 Good relations

In any data environment there are usually several ways in which relations can be constructed and indexed to represent the data and the relationships between the data; but naturally not all of these would be equally 'good'. Some will suit one application rather than another. Some will be more efficient than others and some will include more redundant data than others. One example was discussed earlier, showing that the relation in Table 9.5 is better than the relation in Table 9.4. The choice of relations is obviously vital to the success of any database. However, we leave this discussion until later, to Chapter 11 on database design and specifically to Section 11.3.2 on normalization.

9.2   HIERARCHICAL MODEL

9.2.1. Tree structure

In the Hierarchical model all data are stored in trees. Tree structures were discussed earlier in Sections 7.3.4 and 7.3.5. Much scientific and engineering data can be recorded conveniently and naturally in this form. For example, the data on tensile properties of foil in Table 9.2(a) is already in a hierarchical form and can be represented by the tree in Figure 9.4. The stress data on ferrous alloys in Table 8.2 of the last Chapter can also be naturally represented in a tree as in Figure 9.5.

In general, as illustrated in Figure 9.6, this model supports data in the form of a parent record, with one or more child records, each of which has one or more grandchild records and so on. No child can have two or more parents, but apart from the root node at the top of the tree (the parent record in Figure 9.6), every record has one parent. There are no links across the tree, although sometimes children of one parent are linked together in a chain as in Figure 9.7. Sometimes only twin records, that is, child records of the same type, are chained together in this manner. (The temperature and time records in Figure 9.5 are twins).

Generally a hierarchical structure is stored in one file, the relationship between records being established by their position and type. For example, the general tree structure in Figure 9.6 could be stored in the following serial order, convenient for batch processing:

Parent, Child 1, Grandchild 11, Grandchild 12, Child 2, Grandchild 21, Child 3, Child 4, Grandchild 41, Grandchild 42.

9.2.2 Searches

A hierarchical structure is particularly suitable when all or almost all searches are known to follow a path from the top to the bottom of each tree. Then if the tree is structured using pointers as in Figure 9.6, data at the bottom of the tree can be found quickly. This is the case when we need to find data on the properties of a material. The material (such as type of alloy) and then its condition or state (i.e. temperature, pressure, time, etc) can be represented in successively lower levels of the tree with data on the property at the end of the tree. An illustration of such a tree is given in

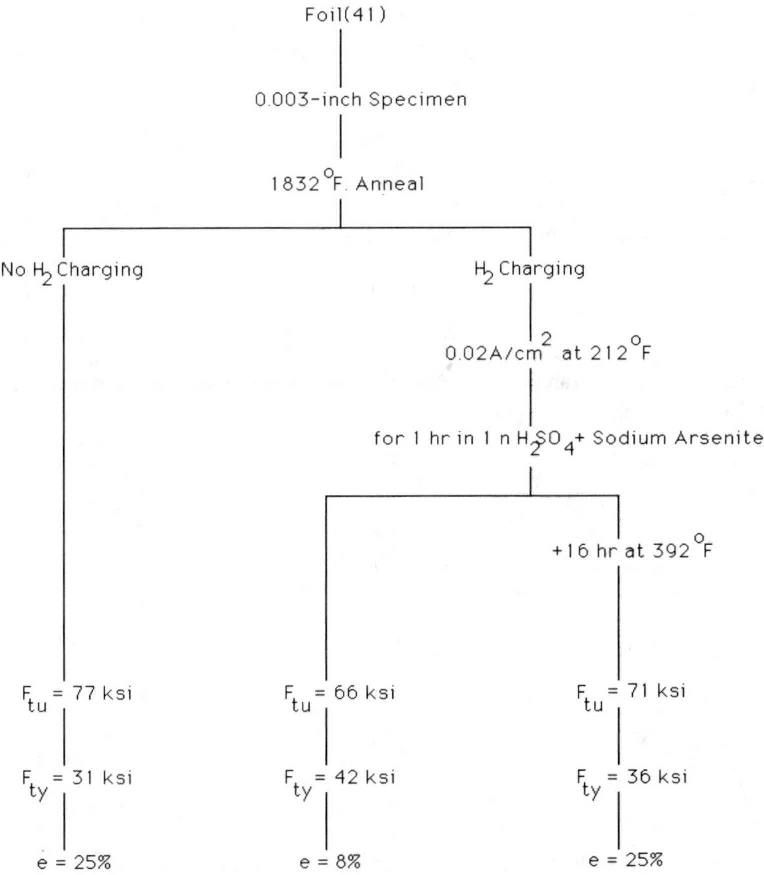

Figure 9.4 Hierarchical structure for the ferrous alloy properties of foil in Tables 9.2(*a*) and 9.2(*b*).

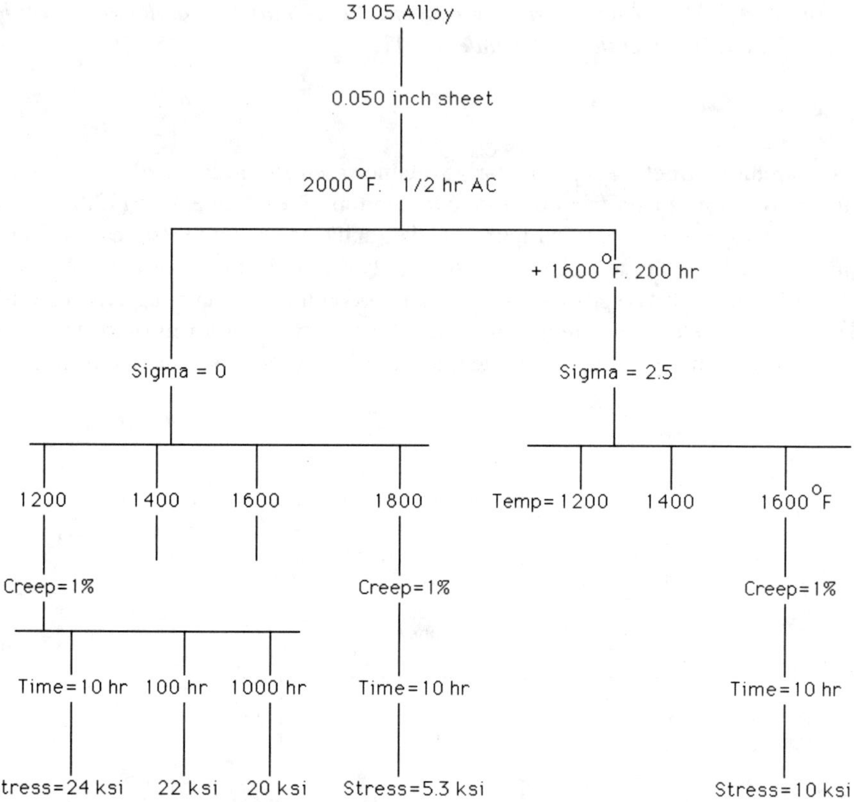

Figure 9.5 Part of the hierarchical structure for data on the sigmatizing treatment of sheet alloy in Table 9.2.

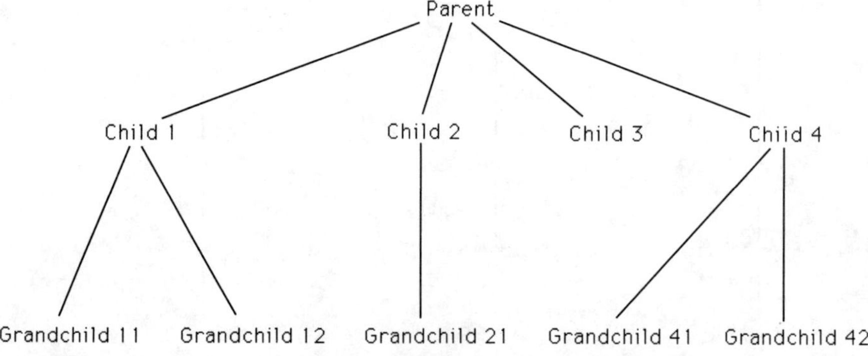

Figure 9.6 General form of tree. Note that all nodes in the tree are linked to one and only one higher level node and to any number of lower level nodes or none.

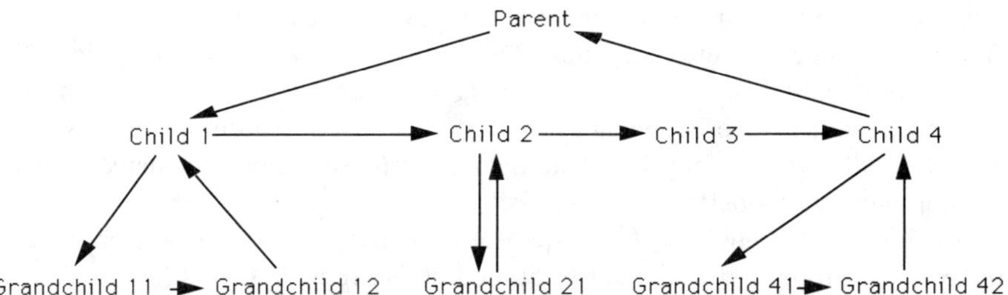

Figure 9.7 Loop structure in which data in Figure 9.4 is often stored. It can also be stored in a serial form (see text) or using lists of pointers from one level to the next as in Figure 9.4.

Figure 9.8. We have added references at the bottom of this tree since scientists normally wish to site the source or sources for each measurement or data item.

A difficulty arises when a search must follow a reverse process, that is when a material is needed with specific numerical values or range of values of some properties, e.g. materials with a tensile strength of 4–5 ksi at 1200 °F. Now all of the trees must be searched from the top material node to the bottom to find which materials match the requirements of the search. Since all of the trees must be searched, this is extremely slow.

Of course, if most requests are for materials with certain specified properties, then trees would be built in the inverse order, with the properties at the top of the tree and

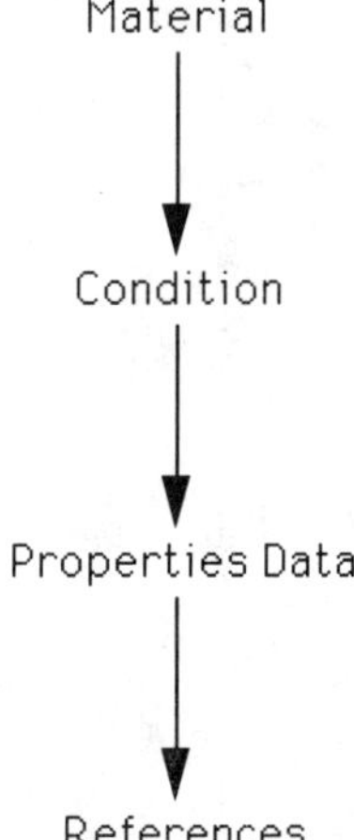

Figure 9.8 Logical representation of a property of materials tree which allows fast retrieval of any property of a material. Each arrow points from a single record to many child records. So the arrows indicate there can be many conditions for each material, many properties for each of these conditions and one or more references for each property data item.

materials at the bottom, i.e. inverted trees with the logical structure in Figure 9.9. But then searches for the properties of a particular material would be slow. So, unless both tree structures are held in the database, which normally would involve an unacceptable level of redundancy, one or other type of search will be very slow. This difficulty is inherent in all hierarchical representations of many-to-many relations such as properties of materials.

An alternative strategy for fast responses to both types of query is to produce an index to the property data at the end of the tree in Figure 9.8.

Let us illustrate this with an example of another similar many-to-many relationship, an ornithological database which holds characteristics of birds in a hierarchical structure as in Figure 9.10. This structure is similar to the one used in almost all bird books, e.g. in the *Audubon Society Field Guide to North American Birds*[5]. The characteristics at the bottom of the tree would consist of such features as:

    Size 35-45'
    Wing span 60'
    Yellow beak
    Blue plumage
    Dark legs

any one of which might be repeated many times as a characteristic of several species.

If you look outside your window and see a bird about 40" long with a yellow beak, blue plumage, dark legs and a black head, you will have to search page after page in the Guide (similar to searching tree after tree in the database) to find the species

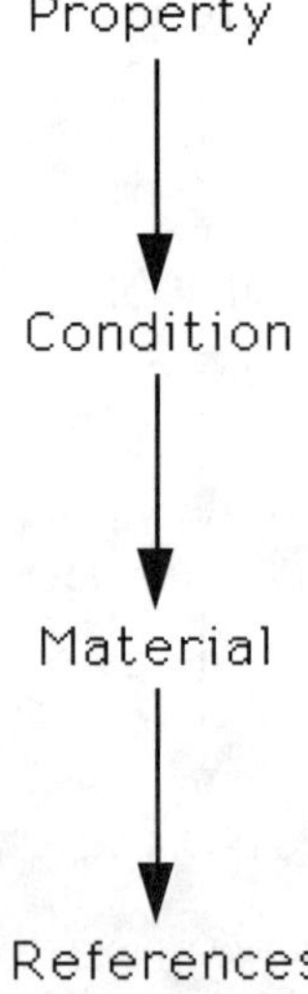

Figure 9.9 Logical representation of a property of materials tree which allows fast retrieval of materials with specified properties, the inverse of the process in Figure 9.8.

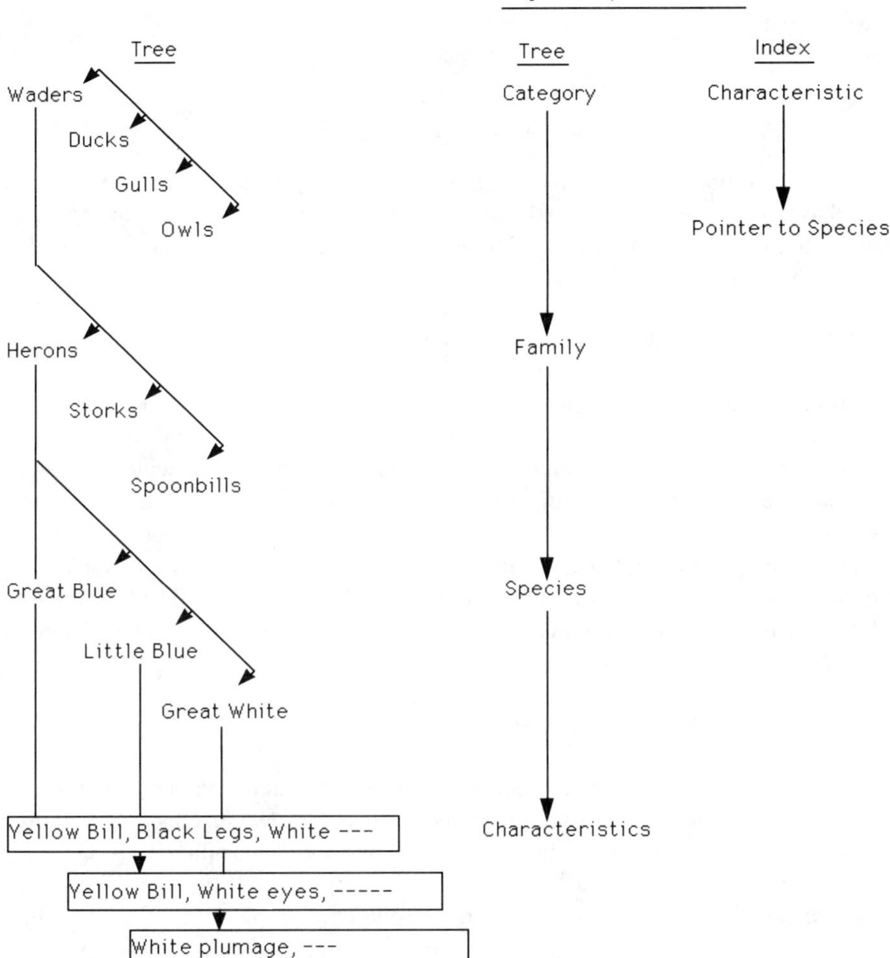

Figure 9.10 Possible hierarchical structure for an ornithological database. An index to speed up queries for species with particular characteristics can take the form of a second tree.

which match the observed characteristics. Some of you may know that this is a slow and frustrating process. By the time you find a species which matches what you saw, the bird will have flown away.

The process would be faster if there was an index at the back of the guide giving a list of page numbers for species with yellow beaks, another list for species with blue plumage and so on, which would direct you to the appropriate pages to inspect. A similar index in a database would consist of a list of characteristics, indexed perhaps with a B-tree (ref. Section 7.3.4) or a hashing algorithm (ref. Section 7.3.5). Stored with each characteristic would be a list of pointers of occurrences of records back in the database. This index could take the form of a second simple tree with two levels:

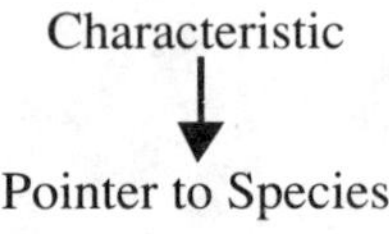

Since this index has the same basic structure as any other file in a hierarchical database this is the method usually adopted in hierarchical databases for speeding up reverse queries, i.e. for finding entities, stored at the top of trees, having particular characteristics or properties, stored at the bottom.

9.2.3 *Hierarchical database management systems*

Although a hierarchical structure is suitable for many technical databases, there are few general hierarchical database systems available. The best known system is IBM's Information Management System (IMS) which hosts a wide variety of physical data structures and indices which make it an effective and efficient system for large corporate databases. It is too large and expensive to be of much interest for all but a few science and engineering applications; descriptions of the system can be found in almost any text on database systems[6]; so we need not describe it further here. Another hierarchical system widely used is System 2000, which is also described in many textbooks[7].

9.2.4. *A hierarchical atomic database*

Although there are few general hierarchical database systems on the market which would be of interest to most scientists, there are some specialized systems in use. One has been built for a database on atomic and molecular physics at Queen's University, Belfast[8].
The database holds numerical data on the following properties:

    Electron ionization of atoms
    Electron excitation of atoms
    Charge exchange of H^+ with atoms
    Interatomic potentials
    Electron ionization of molecules

in a hierarchical structure illustrated in Figure 9.11(*a*). Each tree begins with a property name, so there are five trees rather than one, with numerical values of the properties at the bottom of the tree. This arrangement of a separate tree for each property is convenient because the interim nodes in the trees are not the same for the five atomic properties and because users do not normally want several of the properties of a given atom or ion at the same time.
    It is worth noting here that a structure with a separate tree for each property is often the best structure to adopt, especially in the case when the interim nodes of

trees representing the properties of some objects or materials vary with the property. An example has already been given in Figures 9.4 and 9.5, describing two separate hierarchical structures for two properties of ferrous metals. The nodes describing the condition are quite different in the two cases and two separate trees seem appropriate, both subordinate to a superior root node 'property name' at the top of the tree. This structure would then be similar to the structure for the atomic and molecular database in Figure 9.11(a).

To find data in this atomic database you answer multiple choice questions bringing you down the tree step-by-step, after choosing your units and the form in which you want it printed (e.g. table or graph). An example of a search would be as described in Figure 9.11(b).

Finally, let us look again at the Atomic-Mass and Ionization files. In a hierarchical database these would take the form illustrated in Figure 9.12 which in these simple cases look very similar to three relations linked together. But the relational form is simpler and more convenient. It is this simplicity of the relational model, without the complexity introduced when we use pointers, which makes it so widely used for small and medium sized systems. However, pointers can speed up access to a database, by large factors when the database is large. The next data model we describe also uses pointers and although the structure is complex, it makes it a suitable model for very large databases.

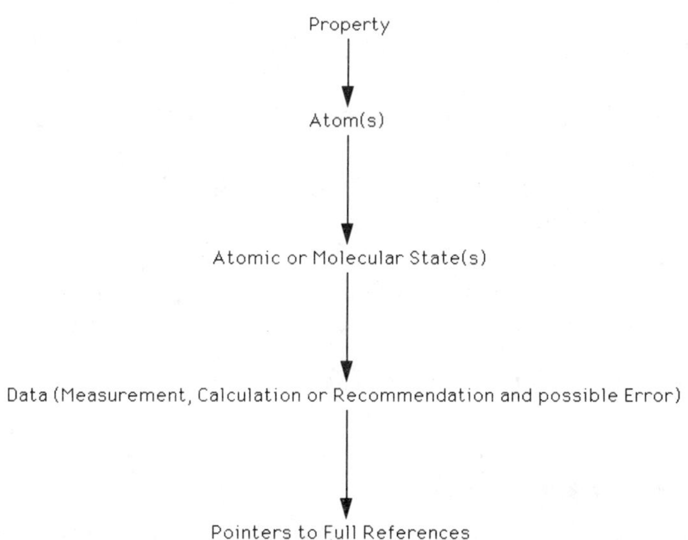

Figure 9.11(*a*) Hierarchical Structure of the Belfast database on atomic and molecular physics (AMDS). Because queries to the database are almost always from the top to the bottom of the structure, this structure suits the user needs, as well as being efficient. Note that there is one tree for each property.

```
AMDS:                   Atomic and Molecular Physics Databases

AMDS:                   Choose process by typing number
                        1. Electron Ionization
                        2. Electron Excitation
                        3. Charge Exchange
                        4. Interatomic potentials
     User:   4

AMDS:                   Type your two atoms
      User:   Li H

AMDS:                   List of states stored for Li H
                        1.  X        1 Sigma
                        2.  B        1 Pi
                        3.  A        1 Sigma
                        99  Any other states
                            Choose
     User:   2

AMDS:                   Energy and length units?
      User:   eV A

AMDS:                   options available
                        1. Particular Experimented Data Sets
                        2. Recommended Data
     User    2

AMDS:                   1. Table, 2. Graph, 3. Curve Fit
      User:   1

AMDS:                   Limits of Range of Distances
      User:   2 5

AMDS:                   Number of points in range
      User:   4

AMDS:                   Potential for Li H, State B 1 Pi
                        R(A°)       V(eV)
                        2.00         - 1.37
                        3.00         - 0.131
                        4.00         - 0.0421
                        5.00         - 0.0128

AMDS:                   1: Table, 2: Graph, 3: Curve Fit
      User:   Q  [Meaning Quit]

AMDS:                   Leaving AMDS. Thank you.
```

Figure 9.11(*b*) Example of dialogue between an experienced user and the hierarchical database system on atomic physics, AMDS, showing how a search progresses down the tree in Figure 9.11(*a*). (The responses of AMDS are modified slightly for the sake of brevity and clarity.)

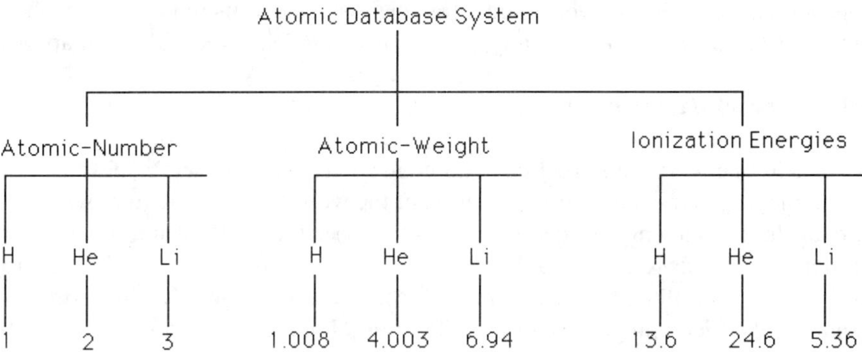

Figure 9.12 *Atom-No, Atom-Wt* and *Atomic-Ionization* files stored in the form of a tree. For simplicity the atomic names have been left out. Note that this is not as simple a means of storing these data as the relations in Table 7.1 and Table 9.4.

9.3 NETWORK MODEL

9.3.1 Background

The Network model, also called the CODASYL model, was developed originally (1969) by the same committee which developed the COBOL language, (the language still used more commonly than any other for commercial data processing). The committee arose from an earlier 'Conference On Data Systems Languages' in 1959, hence the name 'CODASYL' model. However, more often today the name Network model is used.

In Chapter 7 we already mentioned that the Network model has three distinct database languages. These are a schema data description language (DDL), a subschema (user schema or view) data description language and a data manipulation language (DML). The two description languages correspond roughly with the American (ANSI) standards and the original CODASYL DML is being replaced in a similar American standard for a Network Database Language (NDL).

We mention all of this to demonstrate that the Network model is well defined, has international standing and many implementations on several computers. The most well known are Cullinet's IDMS [9] which runs on several mainframe computers (e. g. IBM, ICL) and DEC's DBMS which runs on the larger VAX computers.

However, these systems, like IBM's IMS, are aimed at the large business databases rather than at technical databases. It is interesting that in a DEC manual for their most recent relational database system (RDB)[10], readers were advised to use their network DBMS for large databases with complex relationships. They unfortunately do not mean the kind of complex relationships normally met in science and engineering.

However, research, particularly at the University of Aberdeen, shows that the model can be used effectively in many scientific and medical applications[11, 25].

One advantage in the Network model for scientists is that there is an established FORTRAN interface. Unfortunately it has not been implemented at all installations.

9.3.2 Network data structure

All data in a network data structure is stored in what are called 'sets'. These are not sets in the normal mathematical definition of the word 'set', but are trees with just two levels, one and only one owner record (parent) at the top level and any number of member records (children) at the lower level (including the possibility of zero members). This is illustrated in Figure 9.13. All data must be stored in this form. For example, the hierarchical structure in the ornithological database in Figure 9.10 could be represented by four sets, bird categories, bird families, bird species and bird characteristics as in Figure 9.14. In this example, the Network model is virtually the same as the Hierarchical model. But note that since many species may have any one characteristic, each characteristic may be repeated many times, introducing considerable redundancy to the fourth set. This is because the species–characteristic relationship is really a many-to-many relationship

which cannot be represented directly in the Network model. All data must be expressed in sets which have one-to-many relationships. When we represent the above many-to-many relationship in sets with structure:

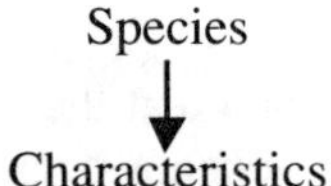

one set for each species, we inevitably allow several characteristics to be repeated many times. In this example the space wasted by the redundancy is not important, but the redundancy imposes slow search times: you still have to search every characteristics record in every species–characteristic set in Figure 9.14 to find birds with a yellow beak.

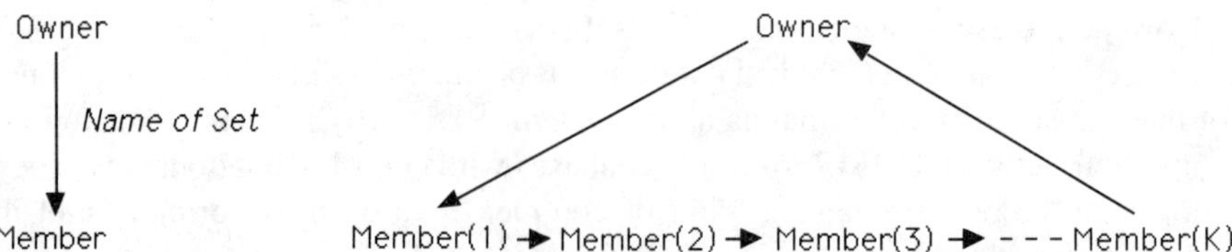

Figure 9.13 Structure of a set with K members in the Network model. The logical form on the left is the one used to describe sets in future diagrams.

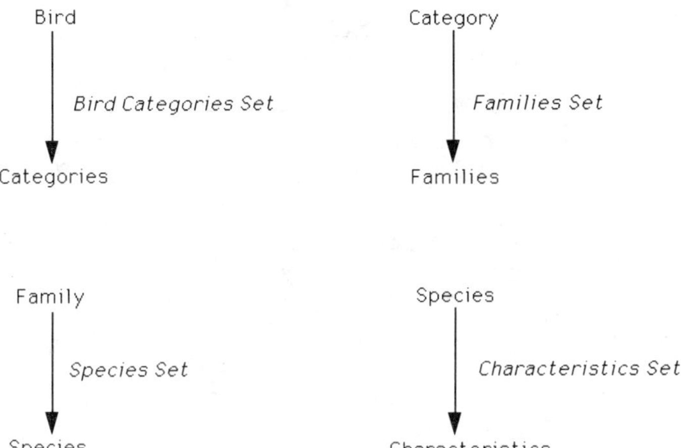

Figure 9.14 Network set representation of the Ornithological database in Figure 9.10 as four sets: Bird Categories, Families, Species and Characteristics sets.

To remove this redundancy in the Network model we introduce a link record between each species and each corresponding characteristic. For example, if we introduce a beak–colour characteristic which has a value yellow, black, red, etc, we can define a link record between this characteristic and each species (e.g. Robin) and include in the link record a value, e.g. 'Yellow'. This link record then becomes a member of two sets, one with parent 'beak–colour characteristic' and the second with 'species' as parent, e.g. with Robin as parent. This is illustrated in Figure 9.15. We have used a second numeric example in this figure, the characteristic wing span of the bird: the value is a number (in inches) or a range of numbers; but the principle of the link record is the name.

This many-to-many relationship is now represented by two sets:

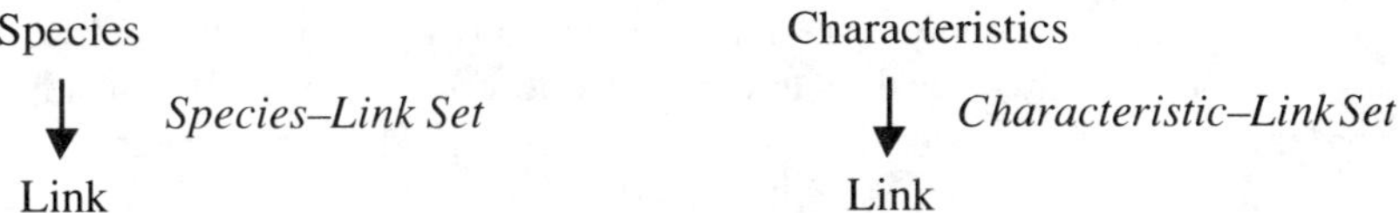

Each link record is a member of the two sets.

This has certainly reduced redundancy to a minimum, but it has unfortunately not much speeded up our search for a bird with a yellow beak because every bird has some colour of beak and we have to search every species–beak link record to find those which are yellow.

So although text books often recommend that a value or quantity be inserted in the link record, in many scientific databases it is best to keep as much information as possible in the owner records. So to speed up searching we might introduce fuller characteristic records: yellow beak, black beak, wing span 0–5', wing span 5–10'

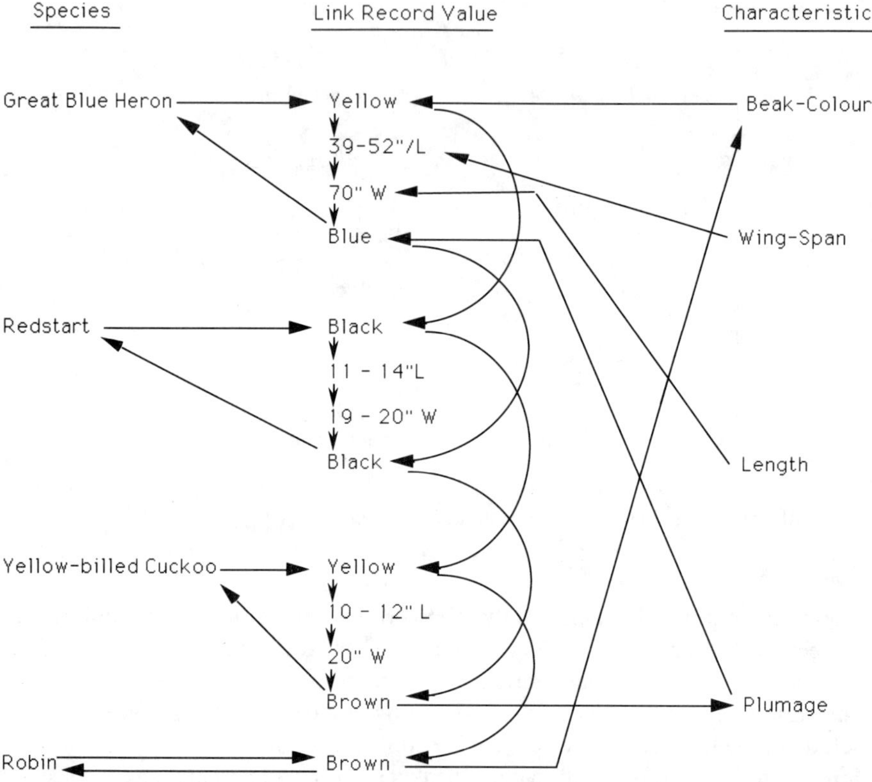

Figure 9.15 Illustration of how in the Network model a many-to-many relationship such as *Species* <<-->> *Characteristics* can be broken into two sets *Species* <-->> *Link* and *Characteristic* <-->> *Link* using Link records containing the value of the characteristics and pointers.

and so on. There are now many more of these characteristic records than before, but this is a small price to pay when the answer to our query on species with yellow beaks is an order of magnitude faster than before and the extra disk space needed is not expensive. The revised structure is illustrated in Figure 9.16.

9.3.3. *Properties of materials*

The hierarchical structure for a property of materials database in Figure 9.8 is easily changed into three sets as in Figure 9.17. There are two changes worth noting.

In the hierarchical structure the condition record may consist of nothing more than one temperature and a set of pointers to data records below. The temperature is concerned with only one material, the one in its parent record; so it is sufficient to store one value only, e.g. 30 °C. But in the equivalent network set

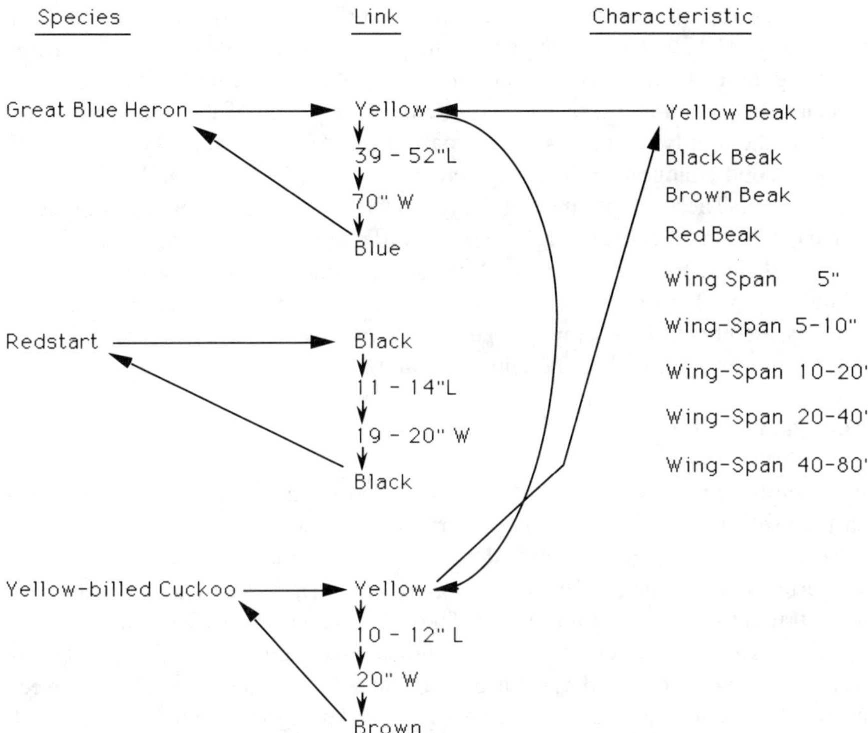

Figure 9.16 By keeping more information on the characteristic in the Characteristic owner record the speed of a search for a species with a particular characteristic can be speeded-up considerably in the Network model. Note that the values, such as 'Yellow', need not be stored in the link record.

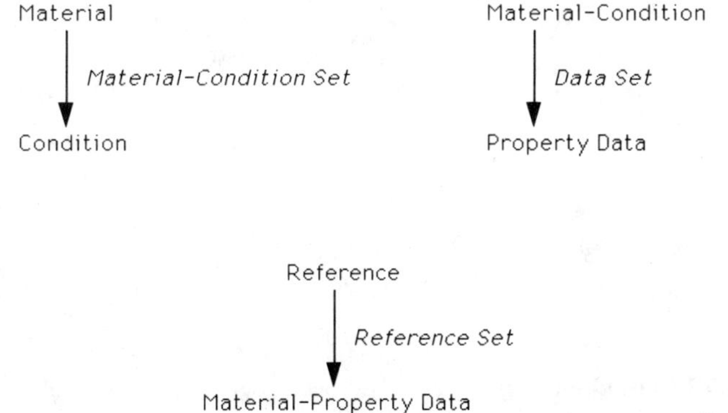

Figure 9.17 Network set structure for a general properties of materials database similar to the logical database in Figure 9.8.

'condition–property', there is no parent to the condition and a condition of 30 °C on its own would be insufficient. So the parent record must include the Material name, i.e. the parent is a Material-Condition record. In the previous Ornithological database this was unnecessary at the intermediate levels of the net because, for example, the family 'Heron' is always part of the 'Waders' category.

The second change from the hierarchical structure is in the last bibliographic reference set. There are often several data items, measurements or calculations, in each paper; so in the Hierarchical model the full reference for the paper would have to be duplicated for each data item. In the Network model we can remove this redundancy by defining a reference set with the reference as parent and the data items as members. This simple example shows how the Network model is, in general, more versatile than the Hierarchical model.

9.3.4 Relational interface

In the implementation of the Network model the records (which must belong to at least one set) are stored in files. The sets are created by adding pointers to these files. For example, in the Ornithological database we would store separate files on categories, families species and characteristics and define sets with pointers. This is illustrated for the set in Figure 9.18. These files can also be viewed as relations. We can add a link between any owner record and its member records by including in each member record the key to its owner. Then during a search the link between the member record and its parent can be established using a join. So it is not difficult

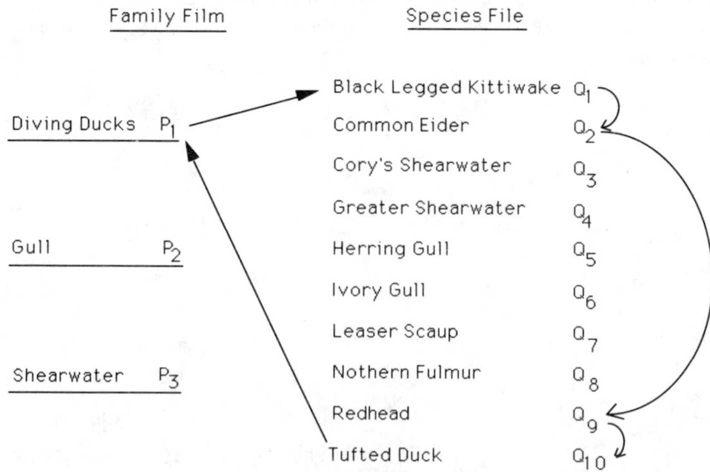

Figure 9.18 The Ornithological Family–Species set structure viewed as two files, a family file and a species file with pointers added to make the sets. Only one set joining four ducks is illustrated. The additional fields P_k in the family file and Q_k in the species file contain other information on the family or species.

to view sets as relations, and because relations are easier to understand, they provide a better man–machine interface. Unfortunately they are also slower.

Let us illustrate with another example. Consider the two physics relations for the alkali metals in Figure 9.19(*a*). These can be stored as a set in Figure 9.19(*b*) in the Network model and in this form many operations are made faster by use of the pointers. For example, to find the resistivity of Na at 40 °C in the Relational model would require the join of the two relations on the common field *Atom-No*. In SQL this would be written:

```
SELECT      R
FROM        Melting-Point, Resistivity
WHERE       Melting-Point.Atom-No = Resistivity.Atom-No.
AND         T = 40 AND Symbol = Na;
```

The join in this operation is usually a slow process compared with navigation using pointers in a network, which in our example we execute using the following steps (slightly simplified):

1. Find the Melting-Point record for Symbol = Na.
2. Find the first Member record of the corresponding Melting–Resistivity set.
3. While T <> 40° find the next Member record
4. When the record for T = 40° is found, then R = 5.5 is the answer.

The above search in the Network model would be completed with only three accesses to the files. The relational computation would normally have taken an order of magnitude more accesses.

Note that if the search had been for a temperature not in the table (say 45 °C) then records for temperatures on either side of 45 °C would have had to be found and the required value of R found by interpolation. But this interpolation process would have been the same in the two models.

Because it is not difficult to view a network as a relation, many network databases now provide the users with a relational interface. Cullinet introduced a special issue of IDMS with this facility, called IDMS/R, but this was apparently so successful that the main product became IDMS/R and the /R was then dropped.

9.3.5 Conclusion

It has been our aim to outline only the main features of the Network model. Most database text books will give good descriptions of the languages and the wide range of facilities available such as privacy and indexing. A short practical description is given by Dinerstein[12] and a fuller description by Deen[13]. Unfortunately most texts base their description in terms of commercial examples (such as accounts or suppliers and parts) which are not typical of most technical databases.

1. Melting Points

Atom-No	Symbol	M.P.($^\circ$C)
3	Li	180
11	Na	98
19	K	64
37	Rb	39
55	Cs	29

2. Resistivity

Atom-No	T($^\circ$C)	R(μOhm-cm)
3	80	11.8
3	90	12.2
3	180	15.0
11	30	5.3
11	40	5.5
11	98	6.9
19	30	7.2
19	40	7.5
19	50	7.8
19	60	8.1
	etc.	

Figure 9.19(*a*) Data on melting points and resistivities of the alkali metals viewed as two relations.

OWNERS			*MEMBERS*	
Melting-point		*Records*	*Resistivity*	*Records*
Atom-No	Symbol	M.P.($^\circ$C)	T($^\circ$C)	R(μOhm-cm)
3	Li	180	80	11.8
			90	12.2
			180	15.0
11	Na	98	30	5.3
			40	5.5
			98	6.9
			30	7.2
19	K	64	40	7.5
			50	7.8
			60	8.1

Figure 9.19(*b*) Data in Figure 9.19(*a*) as it is stored in a Network model.

The Network model is versatile, allowing new relationships between data to be inserted at any time, by the creation of new sets. This permits flexibility and rapid access, but it makes the system complex. Because of the hierarchical structure of much technical data and because of the development of object-oriented databases that include hierarchical structures (i.e. inheritance, see Section 9.6), the navigational approach found in the Network model may grow in importance in the future.

9.4 INVERTED LIST MODEL

9.4.1 Indexed files

Like relational database systems, inverted list systems store all data in the form of tables or files of records, each made up of fields. However, a relation is a **set** of records and therefore has no order; in the inverted list model a file of records is always ordered, usually in alphabetic order or possibly in a hashed order of the primary key. The records can also be indexed with any number of search keys, a search key being any field or combination of fields on which the user wishes to search for a record. Once an index is created, if any insertion or deletion is made to the main file, the index is automatically updated too.

An index consists of a file of the data items which occur in the fields which make up the search key along with the corresponding addresses of the records in the main file. An example is given in Figure 9.20 of our Atomic-Mass file. In this example the pointers are logical record numbers and the system would need also to map the logical number (e.g. record 3) to its physical address on the disk (in terms of blocks

Atomic-Mass File			*A-M Index* File on *Symbol*	
Record No	*Symbol*	*Atom-Wt*	Search Keys	*Address*
1	H	1.01	B	5
2	He	4.00	Be	4
3	Li	6.94	C	6
4	Be	9.01	H	1
5	B	10.82	He	2
6	C	12.01	Li	3
7	N	14.01	N	7
8	O	16.00	O	8
		etc.		

Figure 9.20 Form of index in inverted list structured database representation of a simple *Atomic-Mass* File. The search keys are in alphabetical order to facilitate searching (possibly with a B-tree); but they could be in hashed order.

and bytes). Alternatively (but not likely) the physical addresses could be stored in the index rather than the logical record numbers.

9.4.2 Inverted list

In the example in Figure 9.20 there is always one address per search key. This is because the search key is also the primary key and uniquely defines the record; so there is a 1 to 1 correspondence between the search key and the record number. This is not always the case. The search key can be on a field which is not unique; an example is an Author field in a file of references. Some author's name will occur several times, giving a list of addresses in the index (hence the name Inverted List model). The search key can also be a numerical value or a range of values. An example is given in Figure 9.21 for the electric resistivity of the elements. The index is produced on ranges of resistivities and there are lists of addresses in the index. Because the list of addresses can be very long in many cases, some form of list structure is needed to store them.

9.4.3 Searching

There is no common syntax for searching in an Inverted List database, but there are operations common to all systems.

First there is an operation INDEX. For example, to index on the *Symbol* in the *Atomic-Mass* file in Figure 9.20 we would write an expression such as:

```
INDEX Atomic-Mass ON Symbol TO A-M-Index
```

where *A-M-Index* is the name of the index file (sometimes omitted).

Second there is an operation FIND NEXT which would be expressed in a form such as:

```
IN Atomic-Mass FIND NEXT X := Atom-Wt FOR Symbol = Li USING
    A-M-Index
```

This would find the next record, or initially the first record, in *Atomic-Mass* indexed with the symbol *Li* and assign X to the corresponding *Atom-Wt*.

Operations for searching with ranges of a numerical field, as in Figure 9.21, would be the same in principle. For example, if you were looking for an element with resistivity equal to $10\,\Omega\,\mathrm{cm}$ then the system would check the records for the range 8–16 one by one and find that no element matches the search exactly. An engineer would then want the elements with resistivities as close as possible to 10 units, i.e. Chromium, 12.9 units and Indium, 8.4 units. Although no general database system to our knowledge can do such a proximity search, even though they are so important, particularly in engineering, a user program of a few dozen lines would be sufficient for the computation.

	Electric-Resistivity File			R-Index File	
Record No	Symbol	Element	R(μOhm-cm)	Range of R(μOhm-cm)	List of Addresses
1	Al	Aluminium (99.9%)	2.7	1 1–2	empty 12
2	Sb	Antimony	39.0	2–4	1,6,15
3	As	Arsenic	33.3	4–8	4,11
4	Be	Beryllium	4.1	8–18	10,14,16
5	Bi	Bismuth	107.0	18–32	9,17
6	Ca	Calcium	3.9	> 32	2,3,5,7,8,13
7	C	Carbon	1375.0		
8	Ce	Cerium	75.0		
9	Cs	Cesium	20.0		
10	Cr	Chromium	12.9		
11	Co	Cobalt	6.2		
12	Cu	Copper	1.7		
13	Er	Erbium	107.0		
14	Ga	Gallium	17.4		
15	Au	Gold	2.3		
16	In	Indium	8.4		
17	V	Vanadium	25-26		

Figure 9.21 Example of inverted list index for a file of electrical resistivities where the search keys are for ranges of values.

9.4.4 Join

Now let us illustrate a search involving a join of the *Atomic-Mass* file in Figure 9.20 with the *Electric-Resistivity* file in Figure 9.21, on the common field, the atomic symbol. The search would be for the lightest element (lowest atomic weight) with a high resistivity (> 32).

The algorithm would be as follows (using a simplified syntax for the purpose of illustration):

Minimum-Wt $= 10^6$

```
    REPEAT
          IN    Electric-Resistivity FIND NEXT X: = Symbol FOR
                R > 32 USING R-Index;
```

(This assigns *X* to the symbol in the record corresponding to one of the list of addresses for *R* > 32 in *R-Index*)

```
    REPEAT
        IN     Atomic-Mass FIND NEXT W: = Atom-Wt FOR
               Symbol = X USING A-M-Index;
```

(This assigns *W* to the *Atom-Wt* in the record
corresponding to the symbol *X*. Since there is only 1
value in the list, this is executed only once for each
value of *X*.)

```
               IF W <Minimum-Wt THEN

                   Minimum-Wt: = W;

                   Lightest-Element: = Symbol

               ENDIF;

           UNTIL ENDOF A-M-Index

       UNTIL ENDOF R-Index;
```

The symbol for the lightest element with high resistivity would be in the variable
'*Lightest-Element*'.

9.4.5 Conclusion

The inverted list structure has the main advantage that, like a relational database, it
is easily understood by the user. Retrieval of records can be fast, due to the use of
the index files, but these take a large amount of disk space, often as much space or
even more than the original file. However, disk space is not expensive, and the speed
and flexibility of this model make it attractive.

One problem which occurs in poorly designed systems is the slow time for the
insertion of a new record. Because all of the indices have to be changed as well as
the main file, this can take several seconds. A design which leaves lots of space in
a file (e.g. every third record empty) can largely overcome this, again at the expense
of disk space.

The system is not quite as easy to use as a relational database; but few scientists
or engineers should have difficulties with the simple looping algorithms which must
be written to navigate from file to file, as we illustrate in the above example.
However, many systems now provide relational interfaces and many 'relational'
systems, particularly on small computers, would be more accurately described as
inverted list systems with relational-like interfaces.

The two best known inverted list systems are DATACOM/DB by Applied Data
Research Inc.[14] and ADABAS by Software AG[15], but there are several other
successful systems. ADABAS, although a large expensive system, has been used
successfully on several technical databases; for example, it is used by the IAEA in
Vienna for their large and vital nuclear safeguards database[16].

The inverted list structure is also similar to the inverted file structure used for text
databases (information retrieval systems) described in the next section.

9.5   TEXT MODEL

A free text structure is the most suitable to model the data in those areas of science where the data or information is expressed in English text (or text in some other natural language). Because of the availability and very wide use of word-processors, most information in computer readable form is already recorded as text and this is as true in science as it is in business. This book, journal articles, reports, correspondence, memos and laboratory notes are already mainly in this form. But, as we shall see in Section 9.5.2, the text structure can have a wider application for the storage of some scientific records.

9.5.1 Text files

Text data are normally stored in a text file, the most commonly used file type on PCs and one of the most commonly used on larger computers. A text file consists of an ordered sequence of characters, that is a list of characters, including alphabetic characters, numeric characters and control characters which format the output with indentations, skipped lines, changes-of-font, etc to help enhance the information imparted by the text structure to the final reader. Obviously the order is vital (as it always is in a list). The text file will include delimiters to break a large text into smaller units which contain coherent parts of the information in the whole text.

These parts are often described as documents; sometimes the order of the documents is important when later documents refer to earlier ones. Each document is further subdivided, possibly into lists of sections or chapters, which themselves may be subdivided into subsections, paragraphs, sentences and phrases, but always finally into lists of words. The final delimiters for words are spaces and the various punctuation marks and linefeeds (or carriage return characters). This structure includes not only natural language but also more stylized text, such as the source code of a computer program; this is formatted with indentations, new lines, skipped lines and comments to make it easily understood by a reader and at the same time to make it easily processed using a computer. Another example is discussed in the next section.

9.5.2 Records stored as text

The text file structure is also convenient for the storage of files of records, where each record can be expressed in text form. A record then corresponds to a document. An example is the CODATA referral database which stores records on data centres in science and technology; a typical record is shown in Figure 9.22. Although a few of these fields may contain numbers, they mostly contain descriptions in the form of words. In such cases it is usually more natural, and therefore more user friendly, as well as being more efficient to use a text database than to put these fields into a relation or other structure, designed for more precise and discrete data items. Although most relational database systems allow you to include text in a field (for example to permit you to add a comment), most do not index the contents of the fields or provide the other special text searching facilities provided with a text database management system.

Another example in Figure 9.23 is a record from a museum database, which shows that most of this data is in text form and in Figure 9.24 is a further example, a record in the ornithological database we discussed in Figures 9.14, 9.15 and 9.16. For these kinds of descriptive records, a text database is more natural, much easier for users to understand and more efficient. Other examples are bibliographic records, which are particularly important in science, as most scientists need to keep a list of citations to papers relevant to their work and most data centres need to keep a bibliographic file along with data records. An example is given in Figure 9.25.

```
SERIAL:             (e) de 001

COUNTRY:            Germany FR

MAIN CATEGORY:      Renewable Energy Resources

TITLE:              Solar Informations Centrum c.V. (SIC)
                    Solar Information centre
                    Riedlstrasse 3,D-8000 Munchen 22
                    (Germany FR)

TELEPHONE:          (089) 22-57-55

INSTITUTION TYPE:   Information centre

DIRECTOR:           Hegenbart, R

COVERAGE:           Provision of information on solar energy
                    and energy conservation: energy
                    consulting, public information, setting
                    up data bases.

KEYWORDS:           Climatology; energy conservation; energy
                    economics; solar energy.

OUTPUT:             Publication of printed compilations;
                    Magnetic tapes containing data.

SERVICES:           Provision of specific data upon request;
                    Referral to institutions or published
                    data sources.

AVAILABILITY:       Open to all users; Fees charged.

LANGUAGE:           German; English.
```

Figure 9.22 Example of a record in the CODATA referral database. This is most suitably stored in a text database system (often called an information retrieval system).

```
Accession Card            3

Donor Name & Add          Mr Seamus Kelly
                          1 Ardmore Park South
                          Belfast BT10 OJF

Name                      FLACHTER BLADE

Class                     C4-H

Provenance                Tievebullagh, Co. Antrim

Accession No.             424.1979

Method/Price              Gift

Date                      18.04.79

Day Book No.              637

Accession Description     Blacksmith made iron FLACHTER
                          BLADE. Blade is shield shaped and
                          has pointed tip. The left hand side
                          of the blade has a flat fin shaped
                          wing. Specimen has a long open
                          socket with a single hole and the
                          blade is sharpened on two sides.
                          Measurements: Overall length 47.5
                          cms. Length of socket 14.5 cms.
                          Width at shoulders 40 cms.

Additional Description    Specimen was found on Slieve
                          Tievebullagh, Co. Antrim.
```

Figure 9.23 Example of a record in a museum database. This is another record which is suitable for a text database system. It can also be stored in an extended relational database system which can store and index free text fields, but not as conveniently nor as efficiently.

9.5.3 *Index*

A fundamental part of a text database is the index. Normally there is one index on all of the significant words in the text file. However, in the case of a text file representing a set of records, as in the Codata example in Figure 9.22, there is an alternative: the inverted file may consist of separate indices, one for each field or one for each suitable combination of fields. The easier of these alternatives to under- stand and use, is one based on one universal index. Then to search for the city of Paris in the LOCATION field we need only search for the two words 'Paris' and LOCATION in the one field (or occurring close together), but we can also search for 'Paris' occurring in any field in a record with one instruction.

 The index will consist of two parts: a dictionary of all of the words found in the

```
Great Blue Heron (Ardea herodios)

Habitat:        Common on fresh and salt water.

Range:          Throughout Summer and Winter in South U.S. and
                in Summer in North U.S., Quebec, Nova Scotia.

L:              38"

W:              70"

Description:    Head is largely white, underparts dark and
                speckled, beak rich brown, back slate blue and
                brown. Flies with neck folded.

Alarm Call:     Series of 4 hoarse squawks.

Food:           Fish, Frogs.
```

Figure 9.24 Example of a record in an ornithological database which is more suitable for a text database.

text file, some marked as not significant and therefore not to be indexed or searched, and the others marked as significant. Included in each dictionary record, say for the word 'Paris', there will be a pointer to a list file of references or pointers back to the text file, each pointer addressing a location of the word 'Paris'. This is similar to an index in the Inverted List model in Section 9.4. A diagram illustrating the structure is shown in Figure 9.26.

9.5.4 Text database management system

A text database management system (sometimes called a text information manage-

```
        Title:      Database Systems in Science and Engineering

        Authors:    J. R. Rumble and F.J. Smith

        Publisher:  Adam Hilger Ltd.,
                    Bristol, England.

        Year:       1988

        Abstract:   Most textbooks on database systems describe
                    database systems in terms of commercial and
                    financial applications. This book looks at a
                    database from the point of view of the
                    scientist or engineer, using examples based
                    mainly on technical data in Science and
                    Engineering. Illustrations are given in 3
                    database languages SQL, QUEL and dBASE.
```

Figure 9.25 Example of a bibliographic record, suitable for a text database.

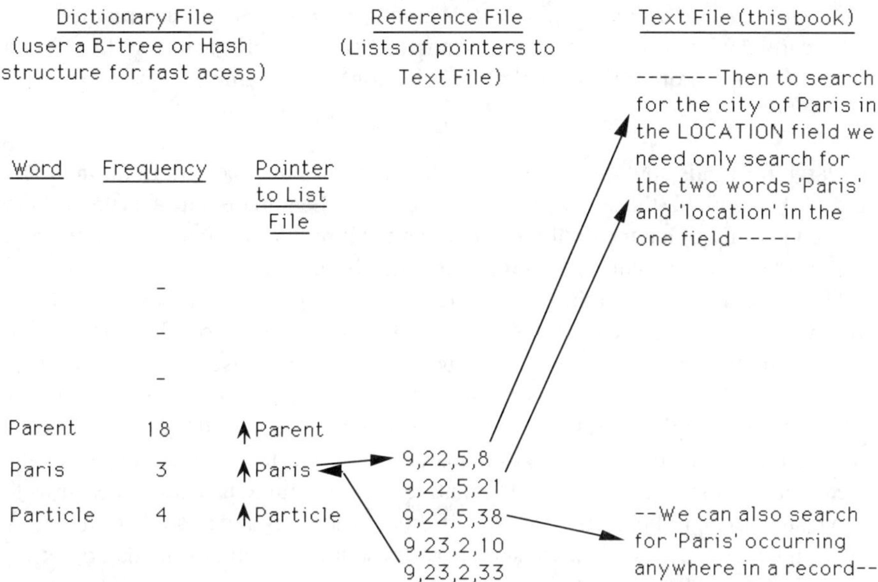

Figure 9.26 Illustration of the inverted file structure used for retrieval in a text database. The first number in a reference file record is the number of pointers to the text file in the record, which may be large for frequent words.

ment system, TIMS) is like any other database system and will consist of the following basic parts:

1. A data description language
2. A data entry program
3. A data editor
4. Utility programs
5. A query language
6. A data manipulation language

1. *The Data Description Language* creates the files for the database and inserts into the dictionary the words not to be indexed (words such as 'the', 'of', 'in', etc.)

2, 3. *Data Entry and Data Editing* are performed using a word-processor or other similar software. After the index is created, the editor should change the index automatically at the same time as changes are made to the text.

4. *Utility Programs* are used for tidying files after numerous changes, modifying the dictionary, re-indexing after a file corruption, giving information on the sizes of files, printing lists of document titles and so on.

5, 6. *The Query Language and Data Manipulation Language* are mainly concerned with searches for words or combination of words using AND, OR and NOT operators (the equivalent of selection). For example,

```
SEARCH FIELD FOR Paris AND Location
```

to search for fields containing both words. Sometimes only certain fields are to be searched or printed (the equivalent of projection). The join operation is unlikely to be needed (although it too could be implemented if required); so the search process is in principle easier than in a conventional database.

The difficulty with searching in free text is that there are a wide range of words with which an author can describe some fact or information. In science and engineering this is less of a problem than (say) in law because technical language is fortunately fairly precise. Nevertheless, it is interesting to note that we could ask for the atomic weight of Helium in four different ways as in Table 9.6. So it is easy to miss important information when searching in free text. Some control of the vocabulary of the text is needed. Alternatively it is useful to provide a thesaurus to a text database to indicate to a user the different synonyms and related words to any word on which the user wishes to search. (A thesaurus is useful in any database; see Sections 2.8 and 8.4).

9.5.5 Information retrieval

A system for searching in a text database is usually called an information retrieval system. There are several systems on the market for mainframes and mini-computers, the best known system being IBM's STAIRS[17], an expensive system which would be suitable for searching in large corporate document databases. On smaller computers, one system designed by one of the authors (FJS) called MicroBIRD[18], is operational on many small personal computers and is limited in capacity only to the size of disk which can be attached to the PC. It can take any text prepared on a word-processor and needs no special editing of the text before the text is input. It can retrieve any document or record in seconds and is being used, for example, for the two sample databases illustrated in Figures 9.22 and 9.23.

Table 9.6 Four ways of expressing the search term in a science example where the language is fairly precise. In other subjects (e.g. Law) the range of possible expressions may be large.

Atomic Mass of Helium
Atomic Weight of Helium
Atomic Mass of He
Atomic Weight of He

9.6 OBJECT-ORIENTED DATABASES

Object-oriented design has recently become a fashionable subject amongst computer scientists as it offers a different approach to traditional software design and it is claimed[19] that it can lead to more efficient programming, particularly because object-oriented code is more readily reusable than conventional code. Object-oriented databases are also the subject of a great deal of research for one additional reason, that they enable us to store, as objects, much more complex representations of the real world than is possible with the other principal models, Relational, Network or Hierarchical.

It is unlikely at the present time that an object-oriented database management system would be used for technical data except for research purposes; the subject is not yet sufficiently developed. However, within a few years, these types of databases will probably become important, particularly in such subjects as computer aided design where the nature of the data which have to be stored and retrieved is more complex than most other data.

To explain object-oriented databases further, we need first to describe the principles of object-oriented programming.

9.6.1 Object-oriented programming

The term **object-oriented programming** has been appearing frequently in the computer literature recently; but the term **object**, as a computer representation of some entity in the real world, is appearing with much greater frequency, and is becoming universal. A record can be viewed as an object. Sometimes a collection of records or a whole file is referred to as an object. However, normally in object-oriented literature the definition of an object goes further than traditional representations of entities as data in programming, because they include not only the attributes of the entity but also the actions or processes associated with the entity. For example, if an object is an atom in an atomic physics database, then amongst the properties might be name, atomic number and atomic mass. The traditional model of an atom might also include multiple occurrences of energy and cross section. However an 'object' goes further and may include processes in its model, including for example:

(i)   the printing of the name and attributes of an atom,

(ii) changing the units in the cross sections from MKS to CGS units,

(iii) the computation of a rate coefficient by integrating the cross sections over a Maxwellian distribution of energies for a particular temperature.

These three possible actions would be stored along with the other data representing the object in the computer system and would become part of the object. This is illustrated in Figure 9.27.

The language describing object-oriented systems is slightly different from traditional systems. The **class** of an object is equivalent to the object type and is defined as a series of **selectors**, similar to fields, which can have two forms:

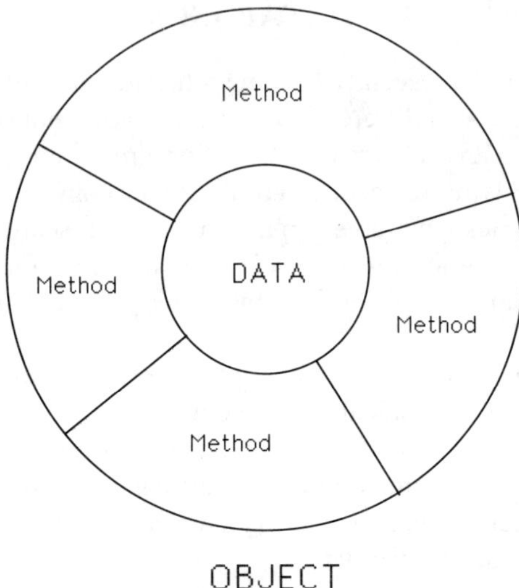

Figure 9.27 Illustration of an object as data surrounded by methods (processes, represented by code) which operate on the data.

(i) **Instance variables**. These are the equivalent of property fields and contain the properties or attributes of an object.

(ii) **Methods**. This is the terminology for action fields and they include the commands to execute actions on the object. In the example above the actions would be 'print', 'change units' and 'compute rate data'.

So **class** is similar to a type and all objects of the same class have the same instance variables (attributes) and methods (processes). So the attribute names, units and methods need only be stored once for a whole class.

The **instance** of a class is a particular object in the class, i.e. the equivalent to a tuple in a relation. One fundamental aspect of object-oriented programming is that when an action has to be performed, such as print, the program need only send a message to the object to execute the command, nothing else, because the object has built into its class the methods needed to execute the actions which can be performed on it. Thus the same action applied to two different classes of objects may have different outcomes. For example the printing of two different classes of objects with different instance variables (attributes) will result in different types of print-outs, i.e. different headings and numbers of columns.

An important feature of all object-oriented programming systems is **inheritance**. In the real world objects are seldom atomistic but are made up of combinations of other existing objects, for example an airplane is a general concept which has a limited number of attributes, such as number of engines, type of engine, weight, range, etc. This will represent one class of object; we can then define new classes (known as **derived classes** or sub-classes) which are defined by inheriting the original class and adding additional properties or actions. For example in the case of our planes, we might classify the planes according to their use or according to the company of manufacture. If we use company of manufacture, then we would add as additional instance variables the name of the company, its address, the factory where the plane was built and so on. There might be a further sub-division, into different types of airplanes, built by that one company, e.g. Boeing 747, 737, 727 and so on. This way a structure is defined for classes, with their associated properties and actions. An example is given in Figure 9.28.

Another feature of some object-oriented systems is called **persistence.** This refers to the persistence of some data in the system after the user is finished with each run of the program. Normally the persistent data are held in files as in any other software system; however in some object-oriented systems it is not necessary for the user to transfer data to files, to make them persistent. This is handled automatically by the object-oriented system.

There are several languages available for object-oriented programming. Three common languages available on a wide range of computers are SMALLTALK, EIFFLE and C++. For the scientist and engineer any of these languages can be used, however C++ is probably the most convenient as it is available on a wide range of computers, including a VAX, and is upwards compatible with C. A good and readable introduction to object-oriented programming has been written by Cox [19].

9.6.2 *Object-oriented databases*

Object-oriented databases view all data as objects. These objects may be any data model of the entities described by the databases; if the objects are all tuples of relations, the database would then be equivalent to a relational database. But

Class	Instance Variables			Methods
Airplane	Fuselage type	Number of engines	Engine type	Draw projection. Print instance variables
Company	Name	Address	Factory built	Print instance variables
Airplane type	Name of type		Year of manufacture	Print instance variables
History	Number of hours flying	Last maintenance date	Parts replaced	Print variables, parts, replaced or not, and dates.

Figure 9.28 Example in object-oriented programming of classes (similar to types), each inheriting the classes above and each with its own instance variables (similar to fields). The actions which can be performed on each class are called 'methods' and are part of the class.

generally we mean by an object-oriented database system one which stores more complex objects, usually one including processes, and one which includes inheritance as a fundamental part of its structure.

The data model may also include operators which can operate on complex objects. For example, a single operation might be sufficient to activate a method of the airplane object which can draw an airplane from a particular angle; it would automatically invoke the operators to draw the parts: body, wings, engines etc. An object-oriented database which includes such complex operations is said to be operational object-oriented, not possible in all object-oriented systems.

9.6.3 Implementation

Object-oriented systems are too new for a universal methodology for implementation to have evolved. Many are designed round a modification of a relational model. One other method is worth mentioning, the use of **frames**.

A frame is not unlike a record or tuple, since all the instance variables and methods (processes) of an object are collected together in a single composite structure. However, unlike the fields of a record, the frame's **slots** (equivalent of fields) are labelled with the attribute names, rather than the normal database convention, to hold the labels in an external schema.

Normalization no longer applies. An instance of the frame may lack any value and since a value is always labelled, if it exists, the slot will not even be recorded. Multi-valued attributes are also permitted.

The storage of processes, represented by explicit code, can have several forms, either in a frame or in some other structure. Source code or semi-compiled code can be stored and interpreted during execution. Alternatively machine code can be stored and linked into the code of a database system during execution[20]. If the code is large, the slot in a frame may contain only an indicator of type and a pointer to the code. The same method would be used for the storage of other large objects such as images or digitally processed signals.

Classes can also be stored in the same manner, in frames. It is sometimes useful to store not only the properties of a class but also default properties of its members.

The above description of object-oriented databases is brief and is meant only to give the reader a general idea of the subject. For further reading we recommend the book by Cox[19] already mentioned and the reports on the International Workshops on Object Oriented Database Systems[21, 22].

We consider object-oriented systems again briefly in Section 12.15 where we discuss object-oriented knowledge-based systems, which may become the main application area of this technology.

9.7 COMPARISON OF MODELS

A survey of database systems on DEC computers[23], which is the range of computers more used by scientists and engineers than any other, found that the underlying data models in the 82 systems examined were as follows:

Relational	40
Hierarchical	9
Network	7
Inverted List	2
Text	6
Other or not known	<u>17</u>
Total	82

All of the recent systems (previous five years) were relational or text systems; but the big installations were still mainly based on the other models.

For data which is not primarily natural language, relational databases are now dominating the market because they are much simpler to understand, both at the design stage and during retrieval.

The difficulty with the Hierarchical and Network models is that the structure is effectively determined by pointers linking files together and must be created by the user before a search is conducted. Unfortunately the correct structure cannot be anticipated beforehand for all applications. In the Relational model any linkage between relations is implemented only by using common data items in fields and executing joins; and therefore the form of this linkage does not have to be decided beforehand. Unfortunately, however, the operations on a relational database, particularly the join, are slow when the database is large and a query in a relational database may be more than an order of magnitude slower than in a Hierarchical or Network model.

Relational interfaces have been implemented on most Hierarchical and Network models, making these systems more readily understood by a wider range of users. But when the data relationships become complex and when a user needs to obtain efficiency from these systems, the underlying structure needs to be understood. Fortunately not all applications require this.

The inverted list structure is so similar to the Relational model that it is relatively easy to add a relational interface. Many of the 40 relational systems, particularly on small micro-computers, are really inverted list systems but give an outward appearance (sometimes not entirely successfully) of being relational. This is an attractive alternative for many scientific or engineering applications.

Text databases are used for a specific type of data which are easily recognized. For more structured, discrete data which can be represented as files of records, a relational system or one of the other three models with a relational interface will normally be preferred.

In many cases, availability, the quality of the implementation and cost are more important factors for the scientist or engineer than the actual model; and whatever system is used we can be sure that some additional special software will be needed to meet the special needs of the scientist or engineer. Object-oriented systems may meet these needs in the future.

REFERENCES

[1] J. D. Ullman, *Principles of Database Systems*, 2nd Ed., Pitman Publ. Co., London, 1982.

[2] J. D. Ullman, *Principles of Database Systems*, 2nd Ed., Pitman Publ. Co., London, Chapter 5, Section 5.2, 1982.

[3] ISO Final text of DIS 9075, Information Processing Systems—Database Language SQL, *Report of TC97/SC21/WG3*, Feb., 1987.

[4] C. J. Date, *An Introduction to Database Systems*, Vol. 1, 4th Ed., Addison-Wesley, Reading, Mass., Chapter 6, 1986.

[5] J. Bull and J. Farrand, *The Audubon Society Field Guide to North American Birds*, A. A. Knopf, New York, 1985.

[6] C. J. Date, *An Introduction to Database Systems*, Vol. 1, 4th Ed., Addison-Wesley, Reading, Mass., Chapter 22, 1986.

[7] A. F. Cardenas, *Data Base Management Systems*, Allyn and Bacon, Boston, 1979.

[8] J. G. Hughes and F. J. Smith, AMDS: A Database System for Atomic and Molecular Physics, *Comp. Phys. Comm.*, **32**, 317–331, 1984.

[9] IDMS Programmer's Reference Guide, Cullinane Corp., Wellesley, Mass., 1978.

[10] Rdb user manual, Digital Equipment Corp., 1986.

[11] P. M. D. Gray and P. Esslemont, *The Performance of a Relational Interface to a Codasyl Database*, Comp. J., 28, 501–7, 1985.

[12] N. T. Dinerstein, *Database and File Management Systems for Microcomputers*, Scott Foresman, New York, 1984.

[13] S. M. Deen, *Fundamentals of Data Base Systems*, Macmillan Press, London, Chapter 5, 1977.

[14] *DATACOM/DB User's Guide*, Insyte Datacom Corp., Dallas, Texas.

[15] *ADABAS Reference Manual*, Software A. G., Reston, Virginia.

[16] Nuclear Safeguards Database, International Atomic Energy Agency, Vienna.

[17] IBM, Storage Information Retrieval System, General Information, Program Product 5734-XR3, 1st Ed., Stuttgart, Germany, 1971.

[18] P. Crookes and F. J. Smith, An Office Document Retrieval System, *Proc. BCS '81 Conference, Information Technology for the Eighties*, 81–92, 1981.
     F. J. Smith and K. Devine, BIRD, QUILL, AND MicroBIRD: A Successful Family of Text Retrieval Systems, Literary and Ling. *Comp.*, **4**, 115–20, 1989.

[19] B. J. Cox, *Object Oriented Programming*, Addison-Wesley, Reading, p274, 1986.

[20] J. G. Hughes, F. J. Hughes and S. R. Tripathy, A Knowledge Base for the Properties of Materials, *Proc. 11th International CODATA Conference*, Ed. P. S. Glaeser, Hemisphere Press, New York, in press, 1990.

[21] K. R. Dittrich and U. Doyal, Eds., *International Workshop on Object Oriented Database Systems*, IEEE Comp. Soc. Press, Washington D.C., p237, 1986.

[22] K. R. Dittrich, *2nd International Workshop on Object Oriented Database Systems*, Springer-Verlag, New York, p373, 1988.

[23] *Hardcopy, The International Guide for DEC Users*, Vol. 6, No. 6, June, 1986

[24] T. W. Olle, Personal Communication, Interim Report of the ISO Committee on the Reference Model of Data Management, ISO/IEC JTC1 SC21 WG3, Database, 1989.

[25] P. M. D. Gray, N. W. Patton, G. J. L. Kemp and J. E. Fothergill, An Object-Oriented Database for Protein Structure Analysis, *Protein Eng.*, **3**, 235–243, 1990.

Chapter 10

Database Planning

10.1 INTRODUCTION

Planning is always the first and necessary prerequisite of a good design. Although it is really part of the design process, we have given it a separate chapter because of its vital importance to the success of any technical database system in science and engineering. The next chapter deals with the remainder of the design process including the entity–relationship diagrams, the choice of model and the choice of relations, trees or sets in the three models.

Too often, because of inadequate planning, technical databases fail or are less successful than they might be. Sometimes they have excellent sophisticated data structures and well-designed software which are the pride of the databases imple-mentor, described at length in many publications—but no satisfied users. Some-times a technical database has some satisfied users; but, because of poor initial planning or because of lack of planning as the database matures, it is not up-to-date with modern technology and is either wasting resources or not providing as good a service as it might. It is worth mentioning here that commercial databases suffer less from these complaints than scientific databases—presumably because of the importance of the systems analyst in commercial data processing and because of the influence of stricter financial control.

Nevertheless, few books on database systems give more than a cursory descrip-tion of the planning process. It is better described in the context of systems analysis; however, books on this subject are not primarily concerned with database systems and help the planner of a database only indirectly. Some useful texts for the planner are those by Rosove[1] who discusses the planning process in many parts of his book on developing information systems; by Davies[2] who discusses information system support for planning in a useful chapter, including selected references; by Murdick[3] who reviews the subject succinctly and ably in a short chapter, including a bibliography; and by Taggart[4] who describes the problems of project planning in one chapter of his book on information systems. One of the most comprehensive texts on the subject is by Blokdijk and Blokdijk[5] who also introduce many advanced topics; but for the average engineer or scientist it may contain too much detail.

These texts show that there are many features contributing to the complexity of a database, all of which must be taken into account during the planning stage. A view of a scientific database system is given in Figure 10.1, showing its inter-related subsystems such as equipment, software, staff, etc, in an environment which includes the data, the user and a source of finance. Each subsystem in this figure is itself complex; for example when considering one of the simplest subsystems, the data, the plan must examine the scope, correctness and reliability of the data, sources of the data, data interchange, security, privacy, cost, conflict with other information sources and, most importantly, the user need for the data and their application. These are illustrated in Figure 10.2.

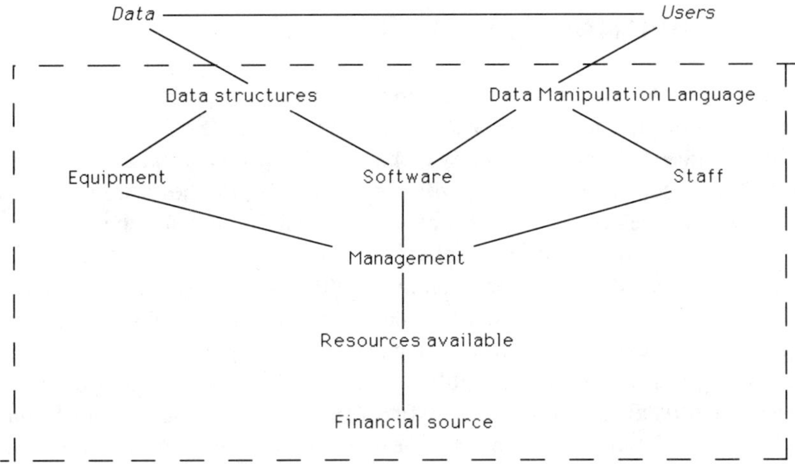

Figure 10.1 View of a database system showing its interrelated subsystems in an environment which includes the data, the users and a source of finance.

Most of these are also complex and can be broken down into other subsystems of lower order. All of these factors are interdependent and most will change with time as better measurements become available and the relations between the numerous factors may also change with time as theory evolves. So the whole process of technical database design, setting up a database, the database system itself, its maintenance and evolution, to which must be added the planning process, has a very high degree of complexity. The planning process is an essential ingredient of this overall system and to be successful it needs a scientific and systematic approach.

10.1.1 Division of problem into parts

The approach suggested in this chapter is based mainly on the reductionist approach to the solution of a complex problem. This is based on scientific principles which

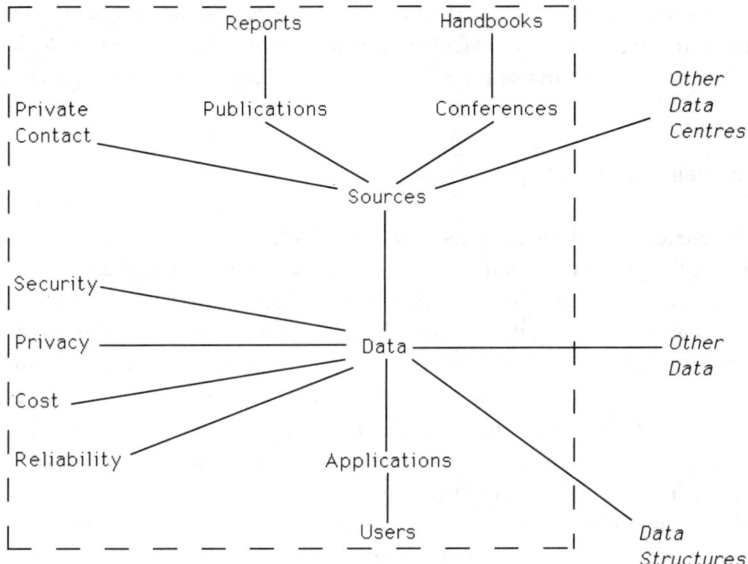

Figure 10.2 Different parts of the data subsystems in Figure 10.1

are not new. A fundamental part of this kind of analysis was described by Descartes in the 18th century: to divide each difficulty 'into as many parts as possible and necessary in order best to solve it'. In other words, each complexity is broken up into component parts and each part examined separately. Each of these parts may also be subsequently broken down into smaller and smaller parts in steps until they are as simple and as well understood as is possible. So the problem of planning a database is changed to one of describing and studying a system, including all of the structures and processes of the database and its management and the relations between them, broken down into parts as small and simple as possible.

It is interesting to note that reductionism is a fundamental part of the systematic approach to the writing of good software known as 'Structured Programming'. The first principle of structured programming, as enunciated by Jackson [6] is as follows:

> *Problems should be developed into hierarchical structures of parts with an accompanying dissection of the program into corresponding structures and parts.*

This is a restatement of the reductionist principle. The same principle is at the basis of any systematic approach to the design of any system just as it is the basis of good planning of a database.

We will say little more about software since it is only part of the overall plan of the database. Care should be taken that during the planning process it does not take up an undue amount of the time and resource allocated. As we have already commented, there will be a natural tendency on the part of a computer scientist, or of many other scientists who are familiar with computers, to concentrate on what they understand and like best, the software and data structures, and give inadequate consideration to other parts of the planning process concerning the users' needs.

10.1.2 Systems approach

To plan those aspects of the database concerned with users, resources and management, where the problems faced are highly complex, a reductionist approach to our planning process by itself is not always sufficient because of the many interdependences between the parts. To help the planning process we can adopt a systems approach in which the database system is examined as a whole in a number of different views[7]. In each of these views the database is represented by a set of subsystems and their relationships, organized to achieve some goal or representing a particular point of view in the environment of the database. Examples are already given in Figure 10.1 and Figure 10.2.

   Another example is the view in Figure 10.3 of a user wanting only to obtain information from the database quickly and inexpensively. Subsystems will then consist of the user terminal, his needs, the output from the system, the cost, etc in the environment of other factors influencing his use (or otherwise) of the database.

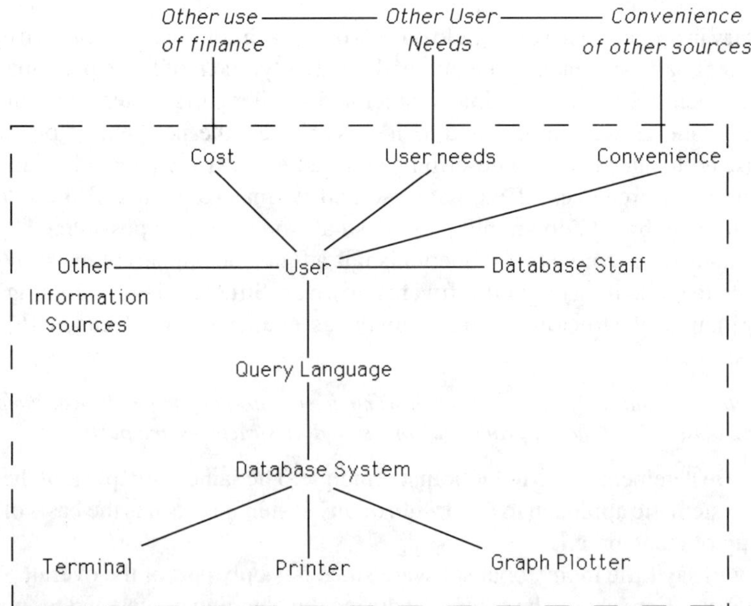

Figure 10.3 View of a database system to illustrate one user's needs in an environment which includes other influences on the user.

The view of the database in Figure 10.3 is particularly important, indeed vital for a successful database. It requires analysis in depth to ensure that the database designer provides all of the data and all of the constituent parts needed by the user community. This will be emphasized again in more detail later.

Another different view of the database may be that of the computer manager with his goal to use his computer resource as conveniently and effectively as possible. The subsystems may now include the magnetic disks, optical disks and tapes, file security, accounting, etc, as in Figure 10.4. Each of these then needs further reduction. Other relevant views are those of the database administration and funding agency; but none are as important as those of the user illustrated briefly in Figure 10.3.

10.1.3 Development cycle

After completing a thorough examination of all relevant views, the systems approach usually proceeds in a number of distinct phases. These depend on the nature of the database system, but they can usually be classified conveniently in four phases:

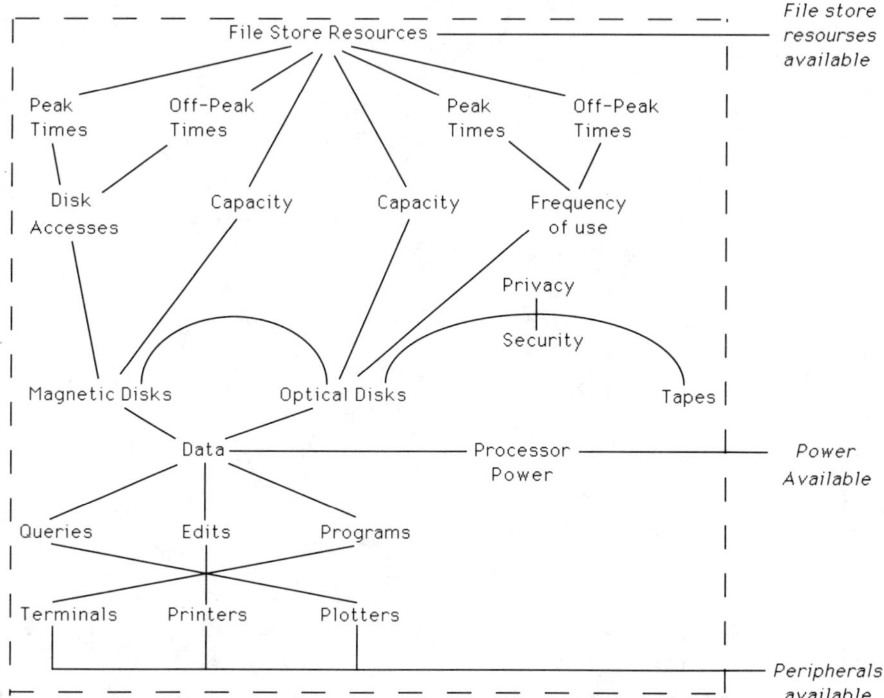

Figure 10.4 View of a database by the computer manager.

    (i) analysis
   (ii) design
  (iii) implementation
  (iv) evaluation.

Each of these for a database system is concerned with

    (i) the data or data structures and
   (ii) the computer system, both hardware and software.

Since the fourth phase evaluation normally leads back to re-analysis and re-design, i.e. back to phase 1 and phase 2, this process is often called the development cycle, as illustrated in Figure 10.5.

The first phase of the cycle, analysis of the whole system, is itself divided into three parts. It begins by determining the objectives of the system, and the nature of the information need. Determining the user needs, including the nature of the data required, is a major task; so it is discussed in greater length later (Section 10.3). When the objectives and user needs are clear, a group of possible alternative existing information systems, database systems or possible expert systems should be examined to see if they can meet these objectives. Modifications may be necessary. A comparison with the costs and effectiveness of a new system should also be

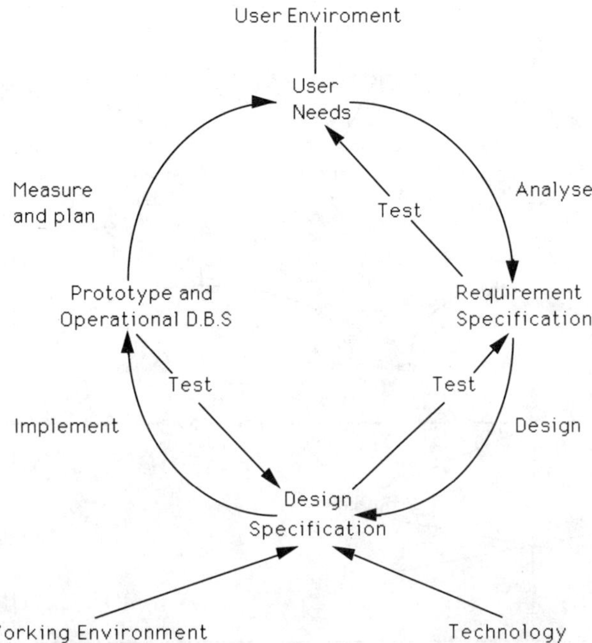

Figure 10.5 Development cycle of database system. The prototype applies to the first few cycles.

studied and the 'best' alternatives chosen to meet the objectives as closely as possible with realistic costs.

The second phase, systems design, will then begin. It is the task of deciding exactly how to implement the steps outlined in the systems analysis study. Entity–Relationship diagrams, choice of data model, choice of relations, trees or sets, data structures and flow charts (or their equivalent in top-down code) are fundamental parts of the design study. These are discussed in detail in the next chapter, Chapter 11. It also involves decisions on data collection, on forms of input and output and graphics.

The third phase, systems implementation, will be initially concerned with the building of a prototype (which is so important that we discuss it separately in the next section), and later, after changes to the plan and design, with the implementation of the database. It is divided into three parts:

    (i)   computer implementation,
   (ii)   data implementation and
  (iii)   organization implementation.

The first of these, computer implementation, is concerned with hardware installation, programming, debugging and testing. The second concerns the capture of data: finding the data, entering data into the system, checking data, correcting data. The third is often just as difficult, but is easily overlooked. It concerns the changes or additions to the organization which is to host the database. It may include training. It may also involve the running of the new system in parallel with any earlier system for a proving period.

The fourth phase, system evaluation, is an assessment of the prototype in the first cycle or two, and of the full project in later cycles after implementation. Even after successful prototypes, full system evaluation often results in changes to the system because of changed or ill-defined objectives.

This completes a brief description of the systems approach as it might be applied to a database system. The principles of the systems approach are those mainly used in the planning process discussed in the rest of the chapter.

10.1.4 Prototyping

Another tool we can use is one fundamental to the scientific approach, one as old as the reductionist approach of Descartes—the use of experiment. Whenever it is possible, decisions on the database or its management should be tested by 'experiments', i.e. by the building of prototypes. Initially these should be simulations and models; the relation of the database with its users should be tested not only by communication with the users but also by some simple tests on the users, e.g., by showing them some sample data in preliminary formats. These testing phases need to be planned too; they should be part of the overall plan[8]. Such testing will therefore be mentioned often in the rest of this chapter.

In addition to these small tests, it is important that at least one large-scale prototype should be built. No matter how careful the planning process is, it is impossible to forsee all of the real needs of a system or to anticipate the reactions or expectations of the users. The prototype will help elucidate both. Expect the reaction: "That's not what we wanted"!

This large-scale prototype, which may take several months to build and analyze, should lead to a re-examination of the whole plan from the beginning. If reasonably successful, the prototype can be modified to become the final product, but a completely new system and possibly a second prototype may be needed. A useful review of research on rapid prototyping, including examples in engineering, has been published in a separate issue of *Computer*[9].

10.1.5 Configuration management

Finally in our search for a systematic and scientific approach to our plan for our database we should not ignore the collection of techniques called 'configuration management' designed originally to improve the quality of hardware products but now also used for software products[10]. This methodology provides greater visibility to management during all aspects of development and production, improving technical and managerial control. Essentially it is concerned with the provision of firm measuring points from which progress on the project can be assessed including such matters as milestones, achievement, quality and documentation. It provides techniques for reviewing, inspection, testing, validation and naming items. Its techniques are well proven and should therefore be made a part of the planning process.

10.1.6 Renewed planning

Planning is needed for the setting up of a database. However, planning does not stop when the database is operational. Renewed planning is needed as a continuous process to cater for major changes in the environment, both in user needs and in technology. However, these changes are often slow, are not immediately visible, and can be overlooked. It is therefore necessary to review the database at regular intervals (say every two years), to plan changes for the future and to review how well the database is working to meet the user needs with current technology. This cyclical planning process has been added to Figure 10.5.

10.1.7 Conclusion

To conclude this introductory section on the planning process, the database planner should begin by reading some of the texts referenced above on the systems approach to solving problems, on systems analysis, on structured programming and on configuration management. A few days spent on such reading should be well worthwhile.

10.2   THE OVERALL PLAN

The setting up of a technical database must arise from some need, usually the need for an information service in some area of science or engineering. Normally the functional requirement is for data to be stored securely in the database in such a way that it can not only provide the information service needed to the prospective users but also allow easy updating and editing of the data by the database staff. It is about the planning of such a database that this chapter is mainly concerned. If the database is of a different type, such as laboratory notebook, analysis, handbook or application as pointed out in Chapter 3, the planning process would have to be altered, but the principles would be unchanged. There are two other possibilities:

(i) a database system may be needed for all users at an installation to store their own scientific data,

(ii) an existing database system may be converting from a manual system to a computer system or from one using old technology to one based on new technology.

The reader should have no difficulty in adapting the following planning process to any of these.

The overall plan for a new database providing an information service is illustrated in Figure 10.6. The process begins with a planner being appointed after the perception of a need for the database. The planner analyzes the user needs (described more fully in Section 10.3) in the context of the constraints of his working environment (described in Section 10.4). He then examines the costs of the possible items of the design and produces a set of alternatives, one of which is chosen and an implementation plan drawn up. These processes are discussed in Sections 10.5 and 10.6 and finally when approval is sought and obtained, provision must be made for planning in the future as discussed in Section 10.7.

10.3   THE USER REQUIREMENTS

The user and his needs should be the most important part of the planning process of most database systems. Unfortunately, as we have previously stated, the necessity to study the user needs is overlooked and the design of the database management system is given too much priority. Just as no commercial product would be designed and built unless a market survey was carried out on its potential sales, so no database system should be designed before the user community and its needs are carefully surveyed.

The process of obtaining information from the user community is illustrated in Figure 10.7 and discussed below.

10.3.1 Data requirement

First, however, the proposed database planner must be clear

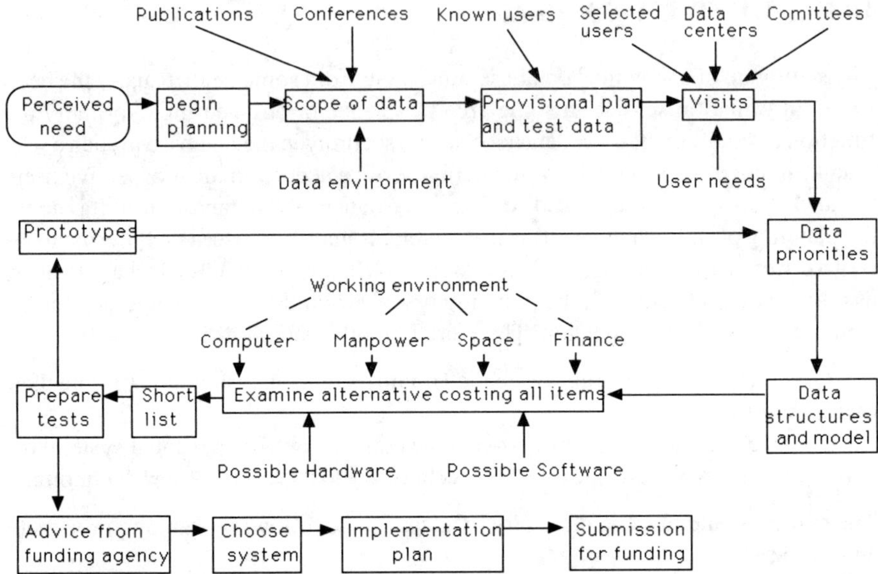

Figure 10.6 Overall plan for a new database to provide an information service in science and engineering

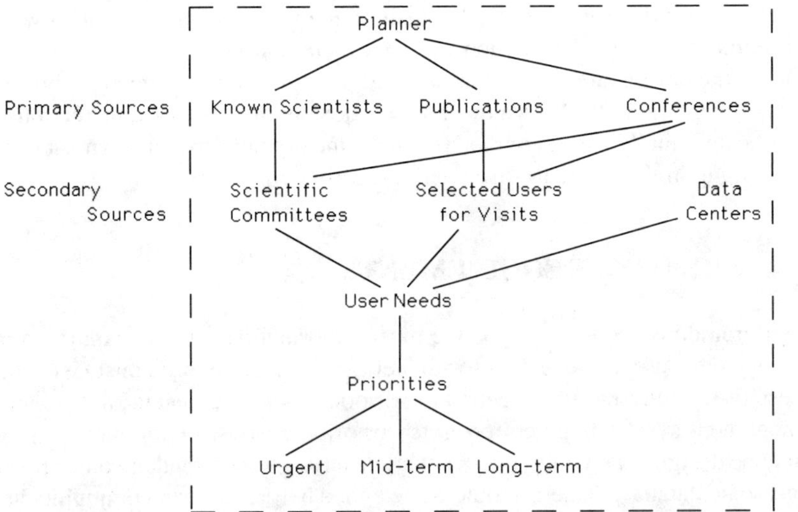

Figure 10.7 Illustration of sources of information on user needs.

(i) what area of data he is considering

(ii) what data he is capable of examining (a physicist could not successfully plan a database on biology)

(iii) what resources are available.

For the planning process to start some need will have been perceived in an area of science or engineering in which the planner has some expertise. He must begin by examining the data in this area in the environment of other related data to define the possible limits on an integrated data system. For example, if the planner has heard that there is a need for an information service on cross sections for electron excitation of atoms he may draw up the view of the data and related data in Figure 10.8. For example, since excitation is always between two energy levels, excitation data are clearly related to data on energy levels. He may find that a good information service on energy levels already exists. He may find that none exists on ionization or wave functions, so these become other possible areas of extension of the database.

On looking at the literature he will find data not only on total cross sections for excitation or ionization but also on differential cross sections and rate coefficients.

10.3.2 *User community*

He must next clearly identify the user community for his proposed data. His original source and local librarian might help. He might find relevant publications, search for other papers citing them and thus find the names of possible users. However, if he does not already know some of the potential users because of his familiarity with the relevant field he should not be considering the database in the first place!

More information can be obtained by correspondence with user scientists known to him, seeking their advice on (i) what data should be collected, (ii) the user community and (iii) relative importance of the data.

Letters to scientists he does not know will have much less value, particularly because of the probable poor response. Scientists generally receive large numbers of circulars, surveys and questionnaires, and they are not likely to give much time to a mailed request for information on a possible new database unless they know the person asking for the information, and at this stage our planner will need advice in depth on the scope of the data and on the proposed user community, not grudging and hurriedly prepared comment.

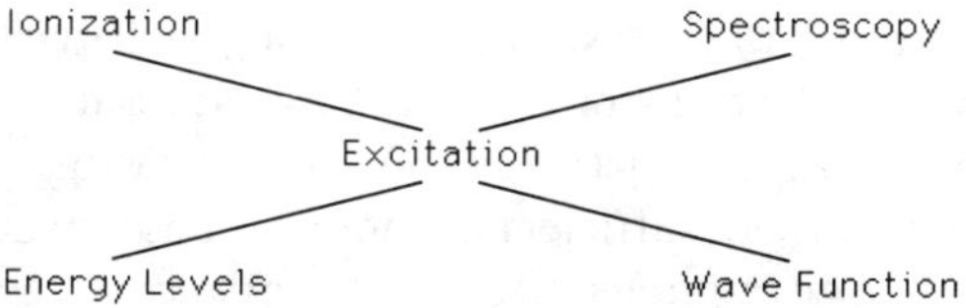

Figure 10.8 Data related to a possible database on electron excitation of atoms. The plan may begin with a database in one data area (excitation data), but discover from a survey of users that the need is for a related area (ionization).

The correspondence could be followed up with telephone calls where the advice given by the correspondents is not clear or when none is received. Approaches could also be made to any national and international committees concerned with the subject by letter or phone.

This should complete the identification of the user community and much information should have been collected on the need for data in the area concerned.

10.3.3 Visit to users

Assuming that this information is positive, the planner should next produce a brief outline of the need for the database and seek support for some travel and for more of his own time to prepare a proposal. Personal contact with some important prospective users, secretaries to appropriate committees and with data centres hosting other databases in the general area cannot be replaced easily by letters or phone calls at this critical stage in the planning process; so travel should be considered essential, even if it is international and has a high cost. Emphasis of the user survey should now be moved to

   (i)   what is thought to be the area of data of greatest importance to the greatest number of users, and
  (ii)   the form in which the users want the data.

Finding the area of data of greatest importance may be vital to the continuation of the project, as it will be on this decision that the first work of the database and the proposal for funding will be based. For example, in the case discussed earlier in relation to Figure 10.8 the user community would be mainly astrophysicists and fusion physicists. The most important area of data for both might be electron ionization, not excitation as planned initially, and the users might want averaged cross sections or rates, not cross sections. So the planner may have started with the aim of collecting excitation cross sections and, if he wants a successful user system supported by the users, end up beginning his database by providing a useful service on rates for electron ionization. If this service is successful, and the users react favourably (by buying his first report or otherwise), it is easier to raise funds for his database to be extended to other areas.

However, the forms in which the data are made available are also very important to the users, if not to the information scientist, who may tend to underestimate their importance and think of them as mere cosmetics. Nevertheless, cosmetics are important when one is trying to attract a user and woo him into becoming an admirer. To find the best format it is not sufficient to approach a user and ask in what form he wants the data displayed or printed; it is highly unlikely that he has given much thought to the problem and he will probably give an inconclusive or misleading answer. It is better to experiment by bringing sample outputs on each visit (they need not contain real data but obviously they would be better if they do). These should show different arrangements, different indexing methods, tabular and graphical forms. This will draw the user's attention and some real useful advice may be obtained. It should then be taken.

In the case of numerical data the user should be asked if he wants the data assessed and authenticated; the answer will nearly always be that he does. He will want error estimates provided in all cases and recommended data by someone he can trust, i.e. by a known expert[12]. Of course source material will always be required. He will prefer a small amount of data of high quality, i.e. which he can trust to be reliable rather than large amounts of unreliable data over a wide area.

10.3.4 Visit to other centres

One warning should be issued about visits to other data centres holding data in related areas. They will almost always have some plans to extend their data collections in other directions and may indicate that they have the data area in the plan almost covered, or about to be covered. A visit and a request to see the actual data may show that the position is very different. If the user survey shows a real need exists which is not being met from an existing database, and provided that the other centre cannot actually show some collected data, the planner should not necessarily abandon his plan. However, it does impose on him the requirement to move quickly to provide some useful data of high quality in the area of greatest need. In today's financial climate this requirement is necessary anyway if funding is to be ensured in the medium and long term.

10.3.5 Storing the user's data

Sometimes a database is designed to enable a group of users to store data, for example, laboratory notebook or laboratory analysis databases (see Chapter 3) for a team working in one laboratory. Then the view of the database by the user for this purpose will be as shown in Figure 10.9. This effectively adds to the system view in Figure 10.3 the additional ability to define the data structure of the user data, to input data, to edit data and to manipulate data. These are not usually required in remote databases, which are providing information services. However, any database will always include the input and editor programs even if only used by the database administrator. For the situation in Figure 10.9, these programs must be made much more secure and much more user friendly before they can be made available to persons other than professional database personnel.

10.3.6 Knowledge rather than data

The analysis of the information requirement of the users may find that the needs cannot be met by data alone. Collections of numbers and alphabetic names may not be sufficient, even if they are structured as tables or linked together in hierarchical or network structures with pointers. For example if we examine any engineering or scientific handbook, we find tables of data on physical coefficients, chemical

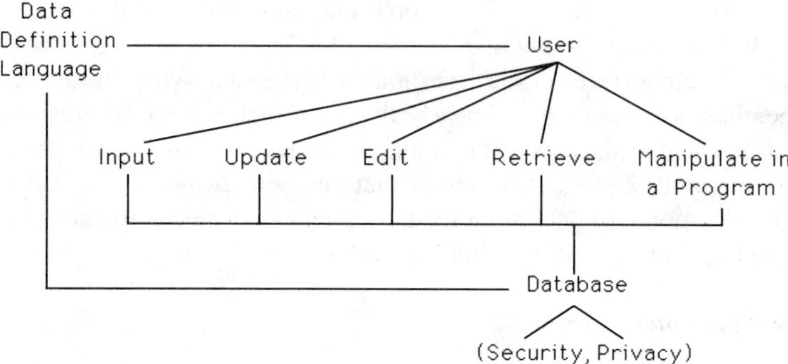

Figure 10.9 User's view of database needed to store user's data.

constants, and so on. However, on almost every page of these tables we find additional information, such as text qualifying the data, formulae explaining how to extrapolate the data or how to compute other data as the physical state changes (perhaps with temperature or pressure). Examples of such tables are given in Table 10.1 and Table 10.2.

It may also include information obtained from experience, for example, a footnote stating that a solvent should not be used in damp conditions. This extra

Table 10.1 Typical example of information obtained in an engineering handbook. The total 'knowledge' expressed below includes a rule for H_v.

| | Heats of Vaporisation† (J g^{-1}) at boiling point | | | | | |
| | Temperature (°C) | | | | | dT/dp |
	0	20	40	60	80	K/Nm^{-2}
Acetone	585	563	541	519	498	0.289
Carbon disulphide	375	367	357	345	331	0.310
Carbon tetrachloride	—	214	208	206	191	0.325
Chloroform	280	273	265	257	248	0.296
Diethyl ether	393	385	374	361	344	0.278
Sulphur dioxide	384	352	—	—	—	0.221

†When the molar volume of the liquid is small compared with the molar volume of the vapour and when the vapour can be treated as an ideal gas the molar heat of vaporization H_v in J mol^{-1} can be calculated from the expression

$$H_v = RT^2 p^{-1} (dT/dp)^{-1}.$$

Table 10.2 Second example where the knowledge content of a table needs comment and a rule in addition to the numeric data.

			Properties of Polymers		
Polymer	P (kg m^{-3})	Tensile Strength (NM m^{-2})	10^5 Coeff Lin Exp (°C)	Heat Capacity (J g^{-1} K^{-1})	Stability
Acetals	1420	65[a]	8	1.46	Attacked by some acids and alkalies good resistance to organic solvents
Cellulose	1480-1530	80-240	—	1.3–1.5	Decomposed by oxide agents and strong acids or by heating to 270 °C. Absorbs water
Cellulose acetate	1220-1340	12-58	8-18	1.26–1.28	Softens at ~130 °C, decomposed by strong acids and alkalies
Chlorinated polyether	1400	39	9[b]	—	Unattacked by acids or alkalis except by oxide acids. Resists most solvents
Phenolic cast resins	1240-1320	35-39	6-8	1.25–1.67	Not affected by acids, little affected by organic solvents

[a]This strength decreases rapidly in the presence of acids approximately in proportion to (P.H./ 7)^{10}

[b]Based on unreliable measurements which may overestimate the coefficient

information, together with the data, constitutes the total knowledge in the tables and a system storing all of it and able to use it dynamically is called a knowledge base system, rather than a database system.

Unfortunately there are no general systems available which make it easy for a scientist or engineer to store the kinds of knowledge which are based on the laws of physics and chemistry, although object-oriented databases may hold the key to this problem in the future (Section 9.6). The only choice available at present is a database system to hold along the parameters of the laws with the data; the laws themselves must be built implicitly into a user program, possibly in FORTRAN.

However when the knowledge can be represented by a set of facts and a set of logical rules, e.g.

IF conditions are damp **THEN** solvent X will be unsuitable

or more generally

IF fact A is true **THEN** fact B has probability P of being true,

these are suitable structures for recording the experience of an expert. Then the knowledge can be represented very effectively using an expert system. Because of their importance to many information problems in science and particularly in engineering, they are discussed later in a special chapter on expert systems and knowledge base systems, Chapter 12.

Often, however, we need a mixture of data and rules. If there are few rules, then a database system can be easily modified by adding user code or comments to be read by the user when the data are displayed, e.g. 'Don't use solvent X in damp conditions'.

If there are a large number of logical rules (say 100 or more), but little data, then an expert system will be more appropriate than a database system. In some cases both might be used, linked together.

At this planning stage all of these possibilities must be investigated and if necessary simulations or prototypes may need to be built in the two environments to compare them. Always be aware that just as a database system with 10 000 records will behave rather differently from one with 100 records, so an expert system with 1 000 rules will be a great deal slower than one with 10 rules.

Since this is a book primarily about database systems we assume in the rest of this chapter and in the next chapter, Chapter 11, that we are dealing with a system with large data requirements and few rules; so we need a database system. Chapter 12 deals with the alternative, when we need a knowledge base system or expert system.

10.3.7 *Data manipulation*

When data are needed from the database for direct input into a user's program, so that the user can manipulate or use the data for some purpose, a data manipulation language (DML) is needed to find the data in the database system (see Sections 8.2.2 and 9.1.5). For scientists or engineers, code from the DML should ideally be incorporated and compiled into their FORTRAN (or other) programs. However, when this is not possible, it is sufficient usually to transfer data to a file which can then be easily read by a FORTRAN program.

It is possible that users may wish to receive data from remote databases in the form of records to be read into their local computer or into their own personal computer for storage and manipulation there. Although such downloading of data will probably not be a user requirement of many users, adequate communications for this purpose should be planned when considering the choice of equipment.

10.3.8 Archiving

Although normally a database is built to provide a service to a user community, occasionally a database is needed only to archive data for some future time; then the plan is simple and must only ensure that no information is lost as data is stored and that it can be retrieved in whatever form is likely to be needed in the future, usually in machine readable form. Most important in this case (and indeed in other cases) is that the file and record structures be well documented and that the documentation be stored in a secure and accessible place (e.g. a library). A copy of the documentation including some sample queries and responses should also be stored with the data. A tape or optical disk copy of the data and documentation should be kept in a separate building for security purposes in case of fire. This last requirement should be a feature of every database.

10.4 WORKING ENVIRONMENT

Having established a need for a technical database system in a particular area of science and technology the constraints (and opportunities) set by the working environment must next be considered, including the staff available (if any), the office space, the available equipment and available database systems. The choice of a specially built database system or use of an existing database management system must be decided. An already existing system is attractive, but it is important to emphasize that the plan should not allow an outdated system to put constraints on the data stored.

In particular, a choice which forces the abandonment of proven user needs should be resisted. For example, inadequate fixed length records should be avoided. Fixed length records may be suitable for commercial applications; for scientific applications they should not be used when they result in loss of information. It is not acceptable to cut off part of a title because it is too long or leave out the fourth and subsequent authors of a paper because the system can only store three. There is no necessity for such restrictions if the system is properly designed: the system should help the database designer, not hinder him.

If restrictions like this exist in the only database and computer available then the planner should look for a more flexible system. Alternatively, if a large database is required with many users, which would need to use a mainframe computer and if a new computer system is planned in a year or two, hopefully with a better database system, the option of collecting the data temporarily on a few microcomputer systems which can later be used as terminals should be considered.

The first choice of equipment for a large database system will normally be the mainframe computer of the organization hosting the database. However, a technical database is usually more suitable run on a small dedicated computer like a MicroVAX if a small number of users need access at one time, or on a large microcomputer or workstation if single user access is sufficient. The speed and storage capacity of these machines is normally sufficient for a technical database, especially with the added use of optical disks.

If a mainframe computer is still the appropriate choice, it should be noted that access for a scientist or engineer definitely needs an intelligent work station, since much processing, particularly graphics, will economically move from the mainframe, where it is expensive, to the work station where it is not. The plan should also use a workstation (which may be a PC) for data input and data validation. It is likely to be found to be both convenient and cost effective.

10.5   THE COST

The overall view of a database shown in Figure 10.1 could also be used with little change to describe the costs, as almost each subsystem in that figure contributes to the cost. However, it is more convenient to look at the costs in the order presented in Table 10.3 where they are broken down into five main headings. We deal with these separately below.

Table 10.3 Breakdown of various items contributing to the cost of the database.

Heading	Contributing Items
Data	Reference Retrieval, Numerical Data Extraction, Consultation, Maintenance
Hardware	Purchase or Rental
Software	Purchase, Rental or In-house
User Liaison	Visits, Conferences, Cooperation, Data Interchange
Management	Personnel, Security, Information Services Insurance, Secretarial, Printing, Overheads

10.5.1 Data

The cost of the data is usually the highest cost of a database, commonly absorbing 50% or more of the budget. It depends on highly paid staff and is therefore increasing and will continue to increase whereas some of the other items in the total cost of the database, particularly computer costs, are decreasing in price. We discuss these costs under four categories.

References

The cost of technical data usually begins with the cost of searching for references. Usually a large number of journals and reports must be searched to find all of the data needed by the users. It is vital that this task be carried out thoroughly as nothing will take away the users' confidence in the database more quickly and finally than the discovery that important data has been left out. This will be helped by using computerized information services now available on almost every subject. Retrospective searches over previous literature can be made using appropriately chosen search terms and a current awareness service can be used to keep the database reference list up-to-date. Any librarian can advise on these.

The cost of such services are small compared with the cost of staff searching in journals in a library with or without the help of abstracting journals. Greater coverage is also obtained from information services since they use much fuller indices of the literature than is possible in an indexing journal. However, this must be checked in the subject area of the database. In the area of atomic collisions we have found that coverage of numerical data items by the main abstracting services retrieves less than 50% of relevant publications[11]. However we believe that this is an exceptional case. Therefore, during the planning process the abstracting services should be checked to ensure that they are adequate and comprehensive. Their cost should be found. The alternative, searching through journals and books in libraries, will be found to be expensive; it will be impossible unless a very good library in the subject area is available.

Numerical data

Normally a technical database will be storing numerical data as well as references; in the case of a scientific paper containing data, it needs to be read and data selected and prepared for data input to the computer system. These then need to be printed and checked, preferably by someone other than the person who prepared the data. This is expensive in manpower. A half man-day may be needed to process one paper; it may be more or less, depending on the nature of the data.

The numerical data to be included may have already been collected and published. In these situations, the cost of locating and entering the data will be less.

Use of experts

The database needs expert staff in the subject to perform these tasks[12]. However, if it is intended to validate the data by the storage of error limits or to store recommended data then it will probably be necessary to employ a consultant, someone who will be trusted by the user community. Good technical consultants are expensive and may charge two or three times the cost of a member of staff per day. Therefore, it is important that all the material and services the consultant needs are well prepared. A visit of the consultant to the centre or a visit of a member of the database staff to the consultant is recommended. Communication with a consultant by mail is not as effective as direct communication; the relatively small cost of travel is usually worthwhile. Depending on the nature of the data, 10% to 20% of staff costs on the database might be spent on consultation.

Maintenance

The last data cost is maintenance. Once the database is established for a year or two, the cost of keeping the database up-to-date will become an increasing part of the overall cost, by replacing old data by newer measurements, by amending recommendations based on the old data and by correcting errors and omissions. Eventually, this will become the main or only activity of the database as it matures. The cost of this must be planned; it is likely to be surprisingly high even after only a year or two. The rate at which data is changing should be determined, for example by a brief examination of the literature over the previous years, and the approximate cost in future years estimated.

This should complete the planned costing of the data in the database.

10.5.2 Hardware

The computer hardware has already been mentioned in the last section with reference to the rapid fall in the cost of computing and the probable use of mini or micro-computers for technical data.

If special equipment is being acquired for the database there are the options to rent or purchase. The advantage of rental, when it is available, is that it reduces the cost in the first year when inevitably the continuation of the database project is being examined for its cost effectiveness. The disadvantage of rental is that it has a very high cost; so it is usually not advisable.

When purchasing equipment three obvious points should be emphasized.

(i) A computer is useless without good software; therefore make sure that all the software used is included in the purchase price, including the operating system and compilers, and check that the software actually is available and working on realistically large files by testing it.

(ii) Good communications with your computer is likely to be important in the future, even if it is not at first. Therefore, make sure that it is possible to connect your computer with another computer made by another manufacturer. If this cannot be done easily, purchase from a different supplier. A minimum capability is an RS232 interface; however, in addition a synchronous interface should be available to an international standard, such as X25.

(iii) If you are buying a small computer with a large disk, make sure that the data can be dumped to tape, floppy disks, optical disk or via a communications link to another computer to guard against corruptions of the data.

If a mainframe computer is being used the cost will usually be made up of several parts such as access time on terminals, mill-time, disk transfers, disk storage and tape storage. Estimates must be made of all of these to find the cost, taking advice from the computer manager.

10.5.3 Software

The main item of software is the database management system. If a DBMS is being purchased or rented for either a large or small computer system, it should be tested with sample data before the agreement is signed. If it needs modification before it can deal with all the requirements of the database, remember that modifications can be long and expensive, whoever is going to carry out the modifications. The sales staff will tend to minimize these difficulties. One way to get an estimate of the difficulty is to ask the sale staff to quote a price and date for the modification. If they cannot give these, be particularly careful.

Be careful also about changing a database system, meant for another purpose, to store technical data. Make tests with some large files as well as small files, to ensure that the system is not very expensive to run. To maximize sales of many database systems the designers include a lot of variable parameters which widen its application. However, they can often only deal efficiently with a small range of these parameters, usually a range corresponding to commercial data (and certainly corresponding to any demonstration). Technical data may be very different and may therefore be extremely inefficient to run.

The simple example we have already given is that much technical information consists of records with a wide variety of lengths which should therefore be stored in variable length records. However, many commercial database systems use only fixed length records; so a length parameter is fixed to enable the largest record to be stored. If it is an author field where 20 or more scientific authors can sometimes occur, the field size might be made large enough to cope with these 20 authors even though such cases are rare. There would obviously be gross wastage of disk space.

Another example known to the authors is a museum which was successfully using a database system for cataloguing typical museum records (number, size, weight, date, place, etc of each item) and tried to use the same system for storing the contents of a set of ancient documents. It was possible—but only after a lot of editing and with a great waste of computer and human resource. A text database system should have been used (see Section 9.5).

10.5.4 In-house system

If a database system is to be written specifically for the database it must be emphasized that programs to input and retrieve data are not sufficient. Programs to edit or amend the data are also needed and these are usually more difficult to write. It may not be clear at first that edits will be necessary almost as soon as the first data are stored.

If a user wishes to use the data from a database in a computer program, the simplest method is to print out the data from the database and then key it back into the program. This is certainly not recommended; besides being slow it is likely to result in error. At the very least a user wants a file of data in the correct format which can then be read directly into his program. But a user will prefer to access the data through the equivalent of a data manipulation language (see Section 8.2.2). Amendments to a FORTRAN compiler to allow such direct manipulation by the user is possible, but would be a large and specialized task.

Lastly, either the editor program or a special program would need to be written to remove old data and to reconstitute the database to remove vacant spaces.

The cost of all of this software is high, at least the equivalent of two man years for a simple database but usually much more. So in almost all cases, an existing database management system which provides all of these facilities should be preferred to in-house software.

10.5.5 *User liaison*

The plan should include continued user liaison after the database project is started. It is vital for success that contact be maintained between the database staff and the user community to ensure that the database is user-driven, not driven by its own needs. This must include a substantial element of travel including short visits to leading users to show progress and seek advice. Visits to conferences are a good way to meet a large number of users at one time.

Co-operation with other data centres should also be planned including occasional visits to ensure that there is not too much overlap in what is being collected and stored. Before long we can expect that data centres will be connected by data communication channels as these drop in cost. Some communication costs should be included in the long term plan to allow data interchange. The first problem of the planner of a new database, however, is to collect some useful data to interchange!

10.5.6 *Management*

Little need be said about management costs except that they must include the cost of secretarial assistance, photocopying and normal office supplies. The photocopying cost of a large number of references may be significant.

The plan must also include the cost of publication of reports. Even if these reports are sold at a profit, the cost of the printing must be met long before any substantial income is available from sales. The cost of printing depends on the nature of the data, the number of graphs and images included. This should be determined from a printer before inclusion in the plan.

Electronic publishing, by providing an on-line service, should also be costed; but do not overestimate the readiness of users to pay for such a service. People prefer books to terminals. The same comment applies to the possible publication of the data on magnetic disks, optical disks or CD-ROMs. You need to be quite certain that your users will accept these possibilities and buy the extra equipment they need to access the service you wish to provide.

Overheads pay for office accommodation, furniture, lighting, heating, air conditioning, information and library services, personnel, security, insurance, etc to the organization hosting the database. There is usually some standard overhead charged by the organization. If not, the cost of the rental of the office space and furniture and all other services must be included.

This would then complete the analysis of the costs of the database.

10.6 ALTERNATIVES

Having discovered the user needs and costed the various alternatives, it is now necessary to draw up a small number of viable options and present them with a recommendation for funding.

10.6.1 Example: plan for a new database

We begin by assuming that the plan is for a new database. The selection of alternatives should start by identifying, say, three areas of data, each larger and including the previous areas which could later become phases of an implementation plan. For example, in our example in atomic physics illustrated in Figure 10.8, we have already discovered that the users' greatest need is for ionization data. Of these data we might have found ionization of hydrogen, carbon, oxygen to be the most important data to the users. So ionization of these could become data area, Phase 1; ionization of the most important 20 atoms could become Phase 2; and ionization and excitation of several atoms become Phase 3. We could then draw up a plan for each phase including data collection, hardware, software, liaison and management, assuming that they would be implemented one after the other.

Within each of these options a number of alternatives should be identified, costed and timed, e.g. purchased software or in-house software, use of organization's mainframe, and many others as illustrated in Figure 10.10. Of these very many alternatives, some will be ruled out quickly for reasons imposed from outside, e.g., insistence from the host organization that its mainframe be used, ruling out a mini-computer option, or the unavailability of rental for some selected computer ruling out the rental option in this case. Many possible computers or terminals or database systems can also be removed from the plan by the lack of good back-up in the locality of the database.

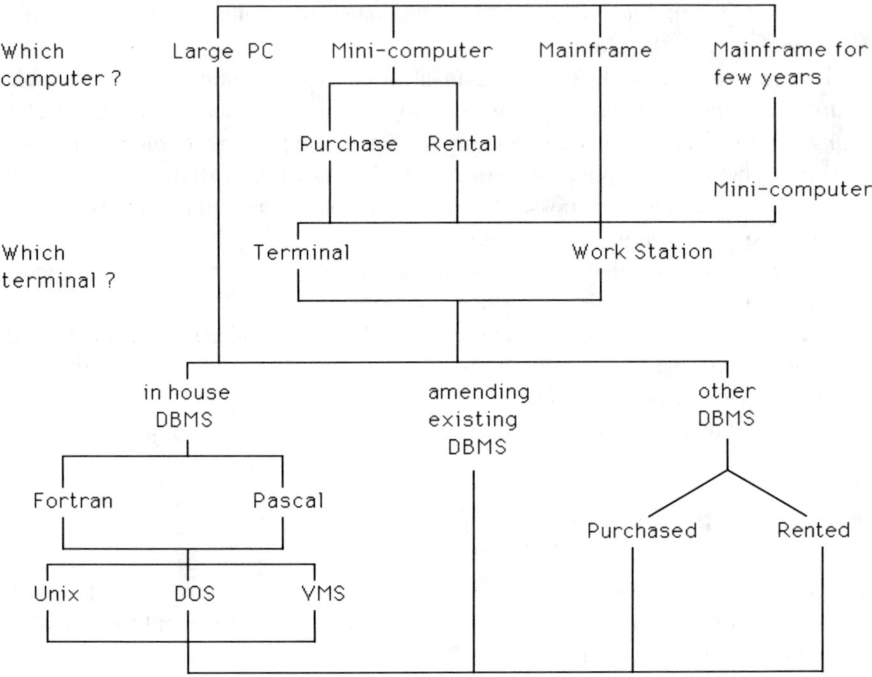

Figure 10.10 Illustration of many alternative hardware/software systems.

From the remaining alternatives a few should be short-listed—one with short term aims, and one with wider and longer term aims. For example we might have two short listed alternatives as follows.

Alternative 1 (short term aim emphasized)

In the first alternative, top priority would be given to the urgent user needs, to meet them as quickly as possible in the first area of data, perhaps within one year. In this alternative, funding might be uncertain after one year; the funding body may view this first year as a trial period to prove the viability of the project, effectively a large prototype. Its success is likely to be judged on the user comments at the end of the year. So it is important to provide useful data, at an early time, for the users, i.e. the data in Phase 1, even if it is to hold up Phases 2 and 3. For example, if we assume that an in-house data system is needed on a mini-computer for Phases 2 and 3 and that this cannot be built in time for Phase 1, then to meet the urgent need of users the data might be compiled with help from a personal computer and small database system (like dBASE) in this first year. Emphasis should be put on quality, not quantity, and on reliability, not haste. Therefore, to accomplish something useful quickly the undertaking would be modest.

Even though the funding body may view the whole first year as a large prototype, it would be wise to build a small prototype and evaluate it, as part of the plan, possibly in the first two or three months of the year's project.

Phase 2 would then be implemented after Phase 1 was finished; it might be completed in (say) three years time.

Phase 3 would not be included in this first alternative plan except to ensure that the database system would be able to handle the additional data in Phase 3 at a later time.

Alternative 2 (mid-term aim emphasized)

This would be like the first alternative but replacing Phase 1 by Phase 2 and Phase 2 by Phase 3.

Thus the first priority would be to provide the users with information on a wider area of data after (say) two years. Work on the database system would be in parallel with the data collection. Work on Phase 3 would start before Phase 2 had finished but with a long term aim of being complete in (say) five years.

For each of these alternatives, the objectives, user advice, staffing, costs and timings should be included in draft implementation plans. The timings might be represented as bar charts, as in Figure 10.11. Tests would be carried out on the proposed equipment and software to ensure that they meet the specifications needed. Simulations might be used to represent the database when it is large to determine response times and costs. Prototypes would be built. These plans and tests could then be discussed informally with the funding agency and final decisions taken on which alternative to recommend for implementation. Obviously the recommendation is more likely to be successful if it is clear and positive, emphasizing the user needs and showing a realistic understanding of costs, cost benefits and (not least) how much finance might be made available.

10.6.2 Documentation

The implementation plan would then be drawn up containing all of the information mentioned above on each alternative along with a clear recommendation.

It is important to emphasise that the plan for a database including alterations be written down, so that all assumptions about service, cost, data included, etc are not subject to misinterpretation. Most database projects involve a wide variety of people, all of whom have different needs, demands and perception. Only written plans are effective in developing a consensus.

The proposal should be summarized at the start with great care, in no more than one or two pages. Those taking the decisions are likely to be very busy people who do not have time to read a long report but who can read a short summary; so those one or two pages are the most important of all. If the need exists, if the funds are

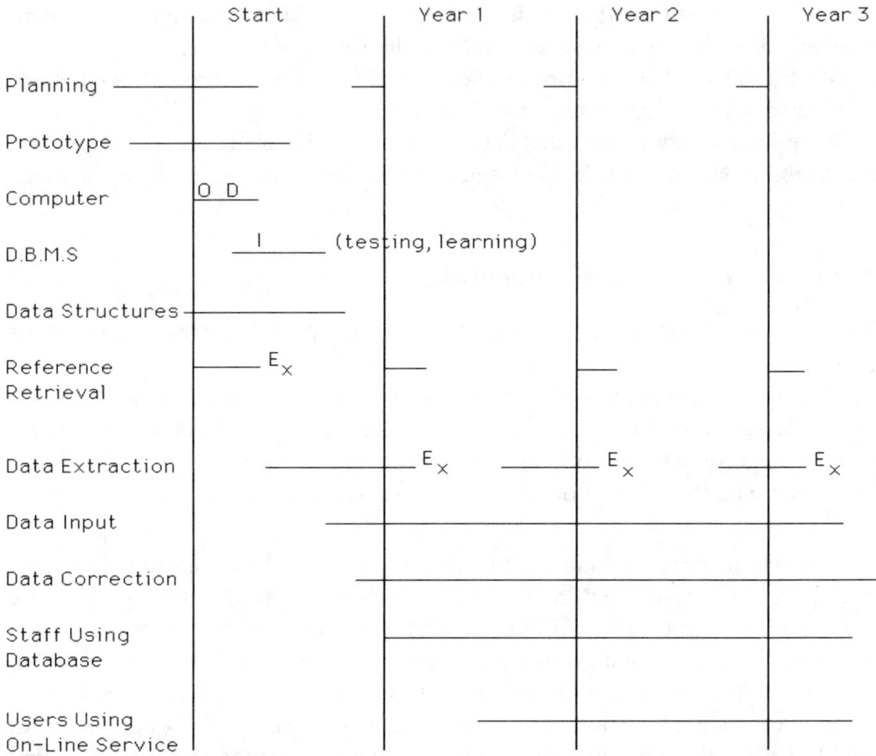

Figure 10.11 Bar charts describing timing of a planned implementation. O:Order, D:Delivery; I:Installation; E_x:Expert.

available and if the proposal is well written, with a little extra luck the proposal will be successful and the project can begin.

10.7   CONTINUOUS PLANNING

Planning does not stop with the implementation of the database but must remain part of the design cycle illustrated in Figure 10.5. Renewed planning is needed as a special process on a regular basis, certainly no less frequent than every two years. Like the overall plan it will begin with a re-examination of objectives, of user reactions to the database and of user needs. Changes in technology must also be reviewed to make sure that the service is as effective and efficient as it can be. Changes in user needs and changes in technology may occur slowly and they may not be immediately visible without a conscious effort to review the system.

A panel of users, meeting (say) once a year, can be a great help to ensure that the database is meeting their needs rather than the needs of the database staff! In addition to the panel the overall plan of the database must include a process of regular review to ensure that the system is aimed at its principle goal—the satisfying of the changing user needs, effective first and efficient second.

10.8 CONCLUSION

As we explain at the beginning of the next chapter on database design, planning can be considered as an essential part of the design process. First, and of overall importance, planning must ensure that the database meets the needs of the users. Second, it must ensure that the database has an effective user interface or it will not be used. Third, it obviously can only achieve these aims if the database is implemented efficiently within the resources available. These principles also must underlie database design, as discussed in Chapter 11.

REFERENCES

[1]   P. E. Rosove, *Developing Computer-based Information Systems*, p384, John Wiley and Sons, Inc., New York, 1967.

[2]   G. B. Davis, *Information System Support for Planning and Control, in Management Information Systems: Conceptual Foundations, Structure and Development*, McGraw Hill Book Co., New York, 338–364, 1974.

[3]   R. G. Murdick, MIS Development Procedures, in *System Analysis Techniques*, Eds. J. D. Cougar and R. W. Knapp, John Wiley & Sons, New York, 87–93, 1974.

[4]   W. M. Taggart, Jr., Project Planning and Management, in *Information Systems*, Allyn and Bacon, Inc., Boston, 284–314, 1980.

[5]   A. Blokdijk and P. Blokdijk, *Planning and Design of Information Systems*, p278, Academic Press, London, 1987.

[6]   M. A. Jackson, *Principles of Program Design*, p299, Academic Press, London, 1975.

[7]   P. Checkland, *Systems Thinking, Systems Practice*, p330, John Wiley & Sons, Chichester, 1981.

[8]   S. Hoffman and S. Inglis, Testing with Databases, *Database J.*, **10**, 13, 1982.

[9]   M. M. Tanck and R. T. Yeh, Eds., Rapid Prototyping: Development Systems of the Future, *Computer*, **22**, No 5, 9–76, 1989.

[10] J. K. Buckle, *Software Configuration Management*, p152, Macmillan Press Ltd., London, 1982.

[11] N. Butterwick and F. J. Smith, *A Study in the use of INSPEC and CHEMABS to Retrieve References on Numerical Data in Atomic Physics*, to be published in 1990.

[12] F. J. Smith and J. G. Hughes, An Expert Driven Database in Atomic and Molecular Physics, in *Proceedings 11th International CODATA Conference, Scientific and Technical Data in a New Era*, P. S. Glaeser, Ed., in press, 1990.

Chapter 11

Database Design

11.1 INTRODUCTION

The design of a technical database system is basically the same as the design of any
other system in engineering. It is based on a systematic analysis of the design
problem and of its possible solutions. It is interesting to look at the general definition
of engineering design by the U.S. Accreditation Board for Engineering and
Technology[1].

> *"Engineering design is the process of devising a system, component or process to meet
> desired needs. It is a decision making process (often iterative) in which the basic
> sciences, mathematics, and engineering sciences are applied to convert resource
> optimally to meet a stated objective. Among the fundamental elements of the design
> process are the establishment of objectives and criteria, synthesis, analysis, construc-
> tion, testing and evaluation. Central to the process are the essential and complemen-
> tary roles of synthesis and analysis."*

Part of the design process as defined above, the 'desired need', and the conversion
of 'resources optimally to meet a stated objective' have already been discussed in
some detail in the previous chapter on planning. We should already have established
the user needs and the financial constraints; we should have before us a plan to
implement a database system in various stages, as outlined in Figure 10.14 and a list
of the queries users are likely to ask or might possibly ask from the database system.
We should know what areas of science or engineering will be initially covered by
the database; so we should know fairly precisely what data is to be collected.

Looking back at the last chapter at Figure 10.5 we are at the point in the
development cycle at the bottom of the diagram entitled Design Specification. In the
cycle this comes just after Requirement Specification, which should now be
complete.

However, the diagram indicates that there is feedback from the design stage to the
specification stage and we can expect that as we develop the design of the database
we will find some, indeed many, questions concerning that specification still
unanswered. We are likely to find that there is still some uncertainty about the data

to be compiled and particularly about the relationships between data. So further contact with users and experts in the field of the data collection will be inevitable as the design unfolds.

We will continue with the design process, under the headings 'synthesis' and 'analysis' with the drawing of what is called an entity–relationship diagram[2,3]. We believe that this diagrammatic approach to database design is appropriate for a scientist or engineer as they are familiar with the use of diagrams. Libraries will have many textbooks on database design, but many will not use entity–relationship diagrams in the design process, relying on a process called normalization described later in Section 11.3.2. We recommend that this be used as a check only, and for technical data to develop the design using the entity–relationship diagram, which should be clearer and more appropriate to the scientist or engineer.

11.2 ENTITY–RELATIONSHIP DIAGRAM

In the last chapter on planning we depicted a systems diagram, Figure 10.8, for one particular database in which the main data areas and the relationships between them are illustrated. Our next step in the design process is to enlarge this diagram into what is called an entity–relationship diagram. The use of entity–relationship diagrams for databases is discussed in many books; a good introduction to the subject is given by Robinson[4]. A different approach is taken by Yourdon[5] and by Elmasri and Noyathe[6]. The last reference includes a useful selected bibliography.

11.2.1 Entities

We take each part of the systems diagram described in section 10.1.2 and write down a list of all the entities, i.e. objects or things, associated with the diagrams and include not only the physical entities but also properties or attributes which describe the entities. We include parts of entities when they might need to be identified (e.g. parts of an engine which have to be ordered—but not arms and legs in a personnel database!) Include also subtypes and superior types when these might be needed. (Subtypes are lower in a hierarchical classification, e.g. subtypes of the entity 'animal' are entities such as 'mammal', 'fish' or 'insect'; the superior type of the entity 'cockroach' is the entity 'insect'.)

To be complete we write down not only all entities and their attributes (i.e. properties) which we know must be included in the database, but also all those which *might* be included and also those which on first sight we do not think should be included. The idea is to include as many entities as possible so that nothing is missed. Without consciously writing down a complete list and then eliminating entities and properties which will not be needed in the database, some entities or properties will be left out. It becomes more difficult to add them again later. Similar advice on writing down all alternatives in the planning stage was given in Section 10.6.2).

One by one all of the potential data items are now considered and we eliminate those which are not relevant to the database system, matching them against the list of likely or possible user queries which we determined at the planning stage. This leaves a final set of data items, i.e. entities and attributes, which will be stored in the database.

11.2.2 Relationships

The data items are next put together in a two-dimensional diagram showing the relationships between the entities and including the attributes. If the diagram is too large then the top-down approach of the systems method, described in the last chapter, should be used to break the full systems diagram for the data down into smaller and smaller units until each can fit on one page (possibly a large page on a drawing board). Each relationship between entities should be marked with a line connecting the entities and a written statement of the relationship should also be written down separately.

11.2.3 School database example

We had intended to illustrate a design of a technical database with an example in science or engineering, such as a database on atomic physics or on tribology. However, to illustrate a design of sufficient complexity to demonstrate all of the points we wished to make would require a detailed and in-depth understanding of both the entities and, what is more difficult, the relationships between the entities in the database and we were forced to the conclusion that only a very limited number of readers would have the detailed knowledge to understand the design in any one technical area. We therefore decided, in this one part of this book, to deviate from our practice of using examples in science and engineering and instead to use an example with which all readers should be familiar—an entity–relationship diagram for a school. The entities are school, teacher, students, age, sex, address, subjects, grades, classrooms, period, year. We assume that students and teachers may change classrooms during the day, that each course, such as Physics in Year 2, is taught by one teacher and that a teacher may teach more than one course. Our database system will aim at answering questions on the following:

1. Who teaches the Physics 2 course?
2. Who is in the Physics 2 course?
3. What courses is student J Barker taking?
4. Where is Mr Jones teaching?
5. What final grade did student J Barker get in Mathematics?

We now examine the entities one by one:

School:	there is only one school so we can leave this out of the database
Teacher:	needed
Student:	needed
Course:	needed
Grades:	needed
Classroom:	needed
Age:	not needed but see Note below
Sex:	not needed but see Note below
Address:	not needed but see Note below
Year:	let us assume that there can be one database for each year and that the database for the previous year is archived at the start of a new year. So *Year* can be excluded, at least initially.

Note. The attributes *Age*, *Sex* and *Address* can help distinguish students and might be useful in the future. So leave these in the diagram for the students; only include the addresses of the teachers.

The entity–relationship diagram would then take the form in Figure 11.1. The relationship must be named and many authors[5,6] recommend that the name of each relationship should be written into the entity–relationship diagram. This makes the diagram cluttered; so we prefer to write down the relationships separately. The relationships written down would include the following:

1. Teachers teach students.
2. Teachers teach in classrooms.
3. Teachers teach courses.
4. Students take courses.
5. Students are taught in classrooms.
6. Courses are taught in classrooms.
7. Students get grades in courses.
8. Teachers give grades to students in courses.

Each of these has one or more inverse relationships, e.g.,

1a. Students are taught by teachers.
8a. Students get grades from teachers in courses.

In addition we have the relationships:

9. Each teacher has an address.
10. Each student has an age.
11. Each student has an address.
12. Each student has a sex.

Of these 12 relations two involve more than two entities, relations 7 and 8. In relation 7 grades are only given to a combination of student and course. So in Figure 11.1

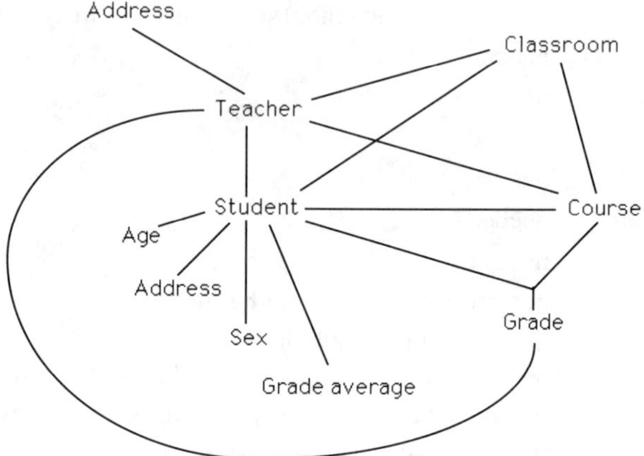

Figure 11.1 Diagram showing all of the entities and relationships in the school database example

we have represented this by a forked relation, the prongs of the fork coming from the entities: student and course.

Similarly, in relation 8 grades are given by teachers to a combination of students and courses; so this is represented by a fork with three prongs. For simplicity, this is represented by one curved line only.

There is one grade which is an exception to the above rules, the overall grade or grade average. This is not dependent on any course, so it differs from other grades and must be entered separately into the diagram with a line joining it to the student

Student ——————— *Grade average.*

So we add a 13th relationship

   13. Each student has a grade average.

11.2.4 Reduction of relationships

An examination of the four relationships 9, 10, 11 and 12 show that these clearly define properties or attributes of the entities teacher and student; so they can be replaced by records with fields:

Student, Age, Address, Sex;
Teacher, Address;

In general each entity in an entity–relationship diagram is represented by a file record. For example, if we expected any questions on the address or size of the classrooms, the classroom entity could be represented by records:

Classroom, Building, Room, Number of seats;

By including attributes or properties which are uniquely linked to one entity as part of the entity the diagram is made simpler.

Before reducing the diagram further we must determine the kind of relationship between entities:

11.2.5 1:1 Relationship

If the relationship of entity X to entity Y is one-to-one, (called a 1:1 relationship) then for each X there is one and only one Y and vice versa. We mark such relationships

$$X \longleftrightarrow Y$$

with a single arrow pointing towards both X and Y.

When the relationship is one-to-one, sometimes X is really a property of Y or Y a property of X. This should be checked. If this is so; the diagram can be simplified by putting X and Y together in one record.

In our example, the only 1:1 relationships in the diagram are the new relation

Student $\longleftrightarrow$ *Grade average*

and the relations involving more than one entity, e.g.

Student, Course $\longleftrightarrow$ *Grade*

Even if X and Y appear to be separate entities, each with its own properties, you should ask the questions

Is the separation of X from Y necessary?

Can X and Y be combined in one record without losing any information and without losing efficiency?

If the answers to these two questions are 'no' to the first question and 'yes' to the second question, then the two records should be combined into one X–Y record. For example, at first the new entity Grade average might have been thought to be a separate entity with a 1:1 relationship with student, but later it becomes clear that the separation of student and grade average is unnecessary and that they can be combined without losing information or efficiency. So grade average would be included in the student record.

11.2.6 1:N Relationship

If the relationship of entity X to entity Y is one-to-many (called a 1:N relationship) then for each X entity there can be many Y entities and for each Y entity there is one and only one X entity. We mark this type of relationship in an entity–relationship diagram

$$X \longleftrightarrow\!\!\!\!\!\gg Y$$

with the double arrow pointing towards the 'many' side of the relationship. Since this is equivalent to a hierarchical relationship, X is often called a parent record and Y is called a child record of the relationship.

For example, in the school diagram the following relationships are one-to-many:

Teacher $\longleftrightarrow\!\!\!\gg$ *Course*
Classroom $\longleftrightarrow\!\!\!\gg$ *Course*

These represent a hierarchical relationship with one parent record and many child records. Sometimes the parent is found on examination to be a property or attribute of the child entity and it can then be included in the child record without loss of information, simplifying the diagram. Even when X and Y appear to be separate entities, if the parent entity is simple, without attributes, then you should ask the same questions we asked in the last section:

Is the separation of X from Y necessary?

Can X and Y be combined in one record without losing any information and without losing efficiency?

If the answers are 'no' and 'yes' respectively then X and Y should be combined.

For example, if we had included the entity Building in the school diagram, then the relationship with classroom would have been:

Building $\longleftrightarrow\!\!\!\gg$ *Classroom*

Then building could have been included in the classroom record provided building was not itself a record. If the building entity had an attribute associated with it, such as address, then by combining building and classroom in one record the address would have had to be repeated many times for each classroom, creating redundancy. So the answer to the second question would have been 'no' and the two entities should stay separate in the diagram.

11.2.7 N:M Relationship

If the relationship of entity X to entity Y is many-to-many (called a $N:M$ relationship) then for each X entity there can be many Y entities and for each Y entity there can be many X entities. We mark this kind of relationship in the diagram:

X $\ll\!\!\!\longleftrightarrow\!\!\!\gg$ Y

In the school diagram the following relationships are $N:M$:

Teacher $\ll\!\!\!\longleftrightarrow\!\!\!\gg$ *Classroom*
Teacher $\ll\!\!\!\longleftrightarrow\!\!\!\gg$ *Student*
Student $\ll\!\!\!\longleftrightarrow\!\!\!\gg$ *Classroom*
Student $\ll\!\!\!\longleftrightarrow\!\!\!\gg$ *Course*

For example, the first relationship states:

Each teacher may teach in many classrooms and each classroom may
have many teachers.

We can write similar statements for the other relations.

Unlike 1:1 relationships and 1:N relationships it is not generally possible to put
the entities at the ends of an $N{:}M$ relationship into one record without causing
considerable redundancy.

So the final entity–relationship diagram at this stage is shown in Figure 11.2.

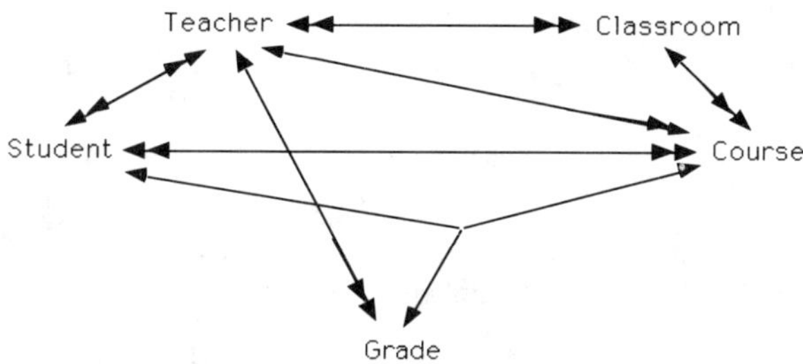

Figure 11.2 Simplified form of the diagram in Figure 6.1 showing 1:1, 1:N and
$N{:}M$ relationships and with attributes now included in the entity records. The
names of the relations, e.g. the name 'teacher' for the *Teacher–Student* relation,
are left out of this diagram to make the diagram simpler.

11.2.8 Removing redundant relationships

There is still considerable redundancy in Figure 11.2. Some of the relations are still
unnecessary, particularly some $N{:}M$ relations. As we explained in Chapter 9, $N{:}M$
relationships are the most difficult relationships to represent in any model. For
example each $N{:}M$ relationship must be removed or modified in both the hierarchical and network models and each causes redundancy in both the relational and
inverted list models. So in as far as it is possible we should try to remove the many-
to-many relations from the entity–relationship diagram.

There are some rules which can help us identify the relationships which might be
removed. We describe them separately. They all involve closed loops of relationships, as we shall see. The removal of these redundant relationships is equivalent
to 'normalization', which is discussed later in Section 11.3.2.

Rule 1. If the two entities X and Y of an $N{:}M$ relationship are both parents of a
common child Z, then the X–Y relationship can often be removed.

This position is illustrated in Figure 11.3(a). If the X–Y relationship is removed
then no information is lost as it can always be reconstructed as follows: for each X
we can find from the X–Z relationship the values of Z corresponding to X and for
each Z we can find a corresponding value of Y using the X–Y relationship. In this way

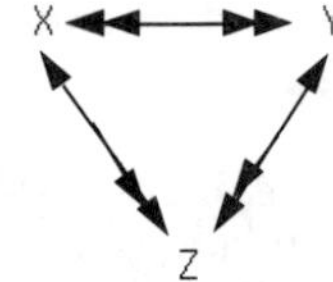

Figure 11.3(*a*)   Situation where two entities X and Y in an N:M relationship have a common child, Z. This X–Y relationship can be removed without losing information.

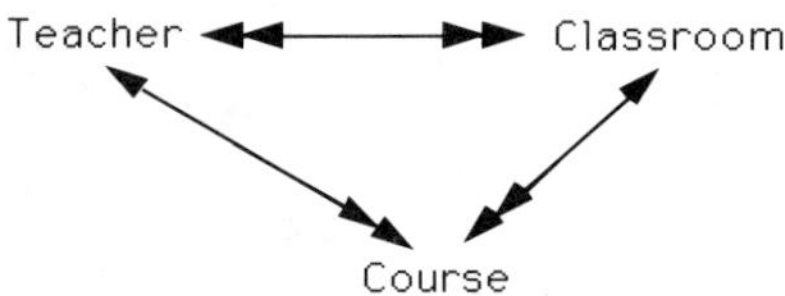

Figure 11.3(*b*) Part of the School database diagram analogous to the situation in Figure 11.3. The *Teacher–Classroom* relationship should be removed.

we can find all Y values that correspond to each X value and vice versa. So the information in the relationship is not lost if it is removed.

When this situation occurs you should ask the following question:

Can the X–Y relationship be removed without a substantial loss in efficiency?

If the answer is 'yes' then the X–Y relationship can be removed. The word 'substantial' is used because if the relationship is not removed, there will be a high cost: for example in the Relational model it will mean an extra redundant relation and in the hierarchical model an extra redundant tree. So it should only be retained if it makes a 'substantial' reduction in efficiency.

Let us look again, by way of example, at Figure 11.2 where we find included in the diagram an example of the above kind of triangular relationship, involving the N:M relationship *Teacher–Classroom* and the entity *Course*. For convenience, this is illustrated separately in Figure 11.3(*b*). We can always find the classrooms in which a teacher holds classes by finding first the courses from the *Teacher–Course* relation and then finding the classrooms from the *Course–Classroom* relation. Also queries based on this N:M relationship, i.e. on classrooms used by a teacher or vice versa, are not going to be very frequent and since no other questions or relationships are affected, we can remove this *Teacher–Classroom* relationship.

Rule 2. The second rule states that when one of two entities, X and Y, in an N:M relationship, say X, has a child Z which makes a second N:M with the other entity, Y, then the X–Y relationship can be removed without losing information. The situation is illustrated in Figure 11.4(*a*) and it is easily shown that the X–Y relationship can be reconstituted from the other two. You should ask the same question as in the first rule:

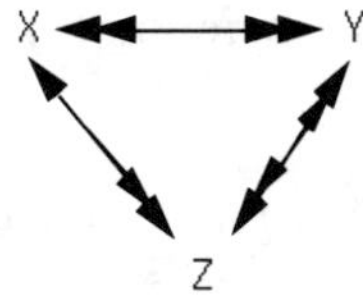

Figure 11.4(*a*) Situation where two entities *X* and *Y* are in an *N:M* relationship and one (*X*) has a child *Z* which has a second *N:M* relationship with the other (*Y*). The *X–Y* relationship can be removed without losing information

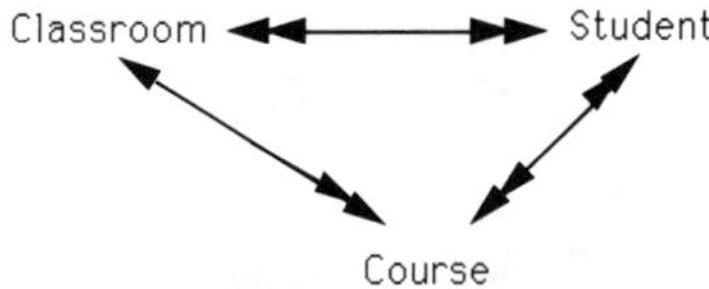

Figure 11.4(*b*) Part of the School database diagram analogous to the situation in Figure 11.5. The *Classroom–Student* relation should be removed

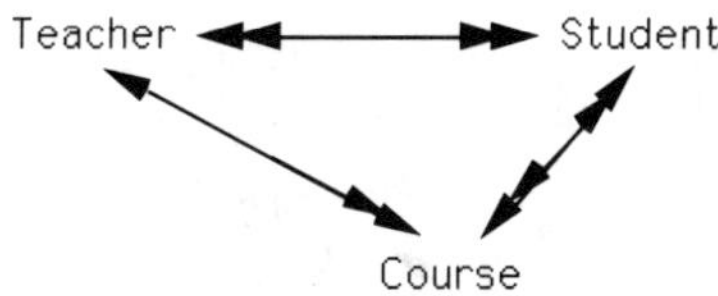

Figure 11.4(*c*) Second example in the School database diagram of a situation similar to Figure 11.5. The *Teacher–Student* relationship should be removed.

Can the *X–Y* relationship be removed without a substantial loss in efficiency?

If the answer is 'yes' then the *X–Y* relationship should be removed.

In the school database diagram in Figure 11.2 we have two examples of this kind of triangular relationship. The first is depicted separately in Figure 11.4(*b*) for the triangular *Classroom–Student–Course* relationship. The *N:M Classroom–Student* relationship is unnecessary, no information is lost by its removal and questions about the classrooms used by students or vice versa are uncommon. So it should be removed.

The second example is illustrated separately in Figure 11.4(*c*) for the triangular *Teacher–Student–Course* relationships. Questions on the students taught by a particular teacher or of the teachers teaching a particular student are not going to be asked often; so the redundant *Teacher–Student* relation should be removed.

Rule 3. If entity *Y* is a child of entity *X* and entity *Z* is a child of *Y* and if in addition a relationship exists, completing the triangle, showing *Z* a child of *X*, then the *X–Z* relationship can often be removed.

This situation is illustrated in Figure 11.5(*a*). Note that the

$$X \longleftrightarrow\!\!\!\!> Y \longleftrightarrow\!\!\!\!> Z$$

relationships form a hierarchy and that no information is lost by removing the

$$X \longleftrightarrow\!\!\!\!> Z$$

relationship. You should then ask the question:

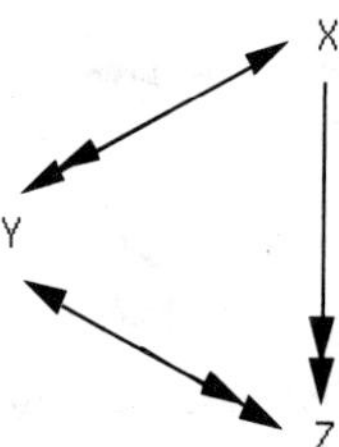

Figure 11.5(*a*) Situation where three activities *X*, *Y* and *Z* are related by three 1:*N* relationships. Then the *X–Z* relationship can be removed without losing information.

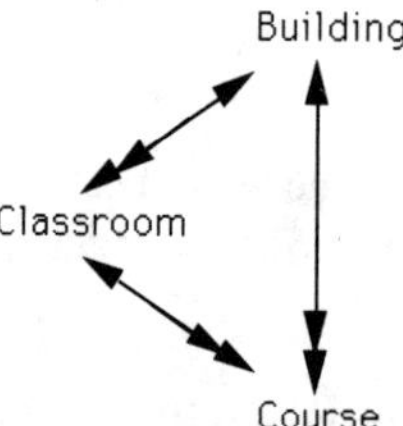

Figure 11.5(*b*) Part of a School database diagram which includes the entity *Building* which is analogous to the situation in Figure 11.5(*a*). The *Building–Course* relationship should be removed.

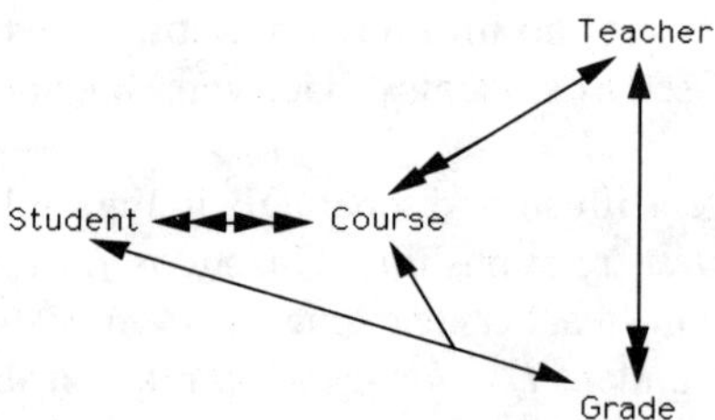

Figure 11.5(*c*) Another similar situation to Figure 11.5(*a*) for the School Database, though not as clear as Figure 11.5(*b*). The *Teacher–Grade* relationship can be removed.

Can the *X–Z* relationship be removed without substantial loss of efficiency?

If the answer is 'yes' then it should be removed.

There is no example of this in the school database illustrated in Figure 11.2, but if we had included a building record with two additional relationships:

A class room is in a building (*Building* <—>> *Classroom*).
Courses are taught in buildings (*Building* <—>> *Course*).

which are illustrated in Figure 11.5(*b*) then the *Building–Course* relationship could be removed without loss of efficiency because few questions on the building used by courses are likely.

Rule 3 Extension. Note that Rule 3 also applies when entity *Y* is replaced by a string of entities and relationships, as long as they represent a tree structure. So this rule tells us that in all of the examples illustrated in Figure 11.6 the *X–Z* relationship can be removed, provided it does not result in a substantial loss in efficiency.

However in the situations in Figure 11.7 the *X–Z* relationship cannot be removed as it cannot be reconstructed from the other relations.

Rule 4. If entities *X* and *Y* have an *N:M* relationship and if both have a common parent *Z* then either the *X–Z* or the *Y–Z* relationship can be removed without losing information as either can be reconstructed from the other two relationships.

The situation is illustrated in Figure 11.8(*a*). You must find which of these two 1:*N* relations is least likely to be involved in user questions and then ask the question

Can this relationship be removed without substantial loss in efficiency?

If the answer is 'yes' then it should be removed.

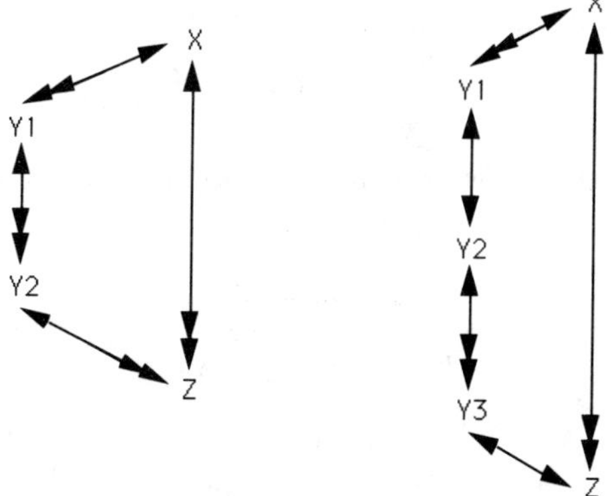

Figure 11.6 Situations similar to that in Figure 11.5(*a*), where the *X–Z* relationship can be removed without loss of information as the *X–Z* relationship can always be reconstructed from the other relationship.

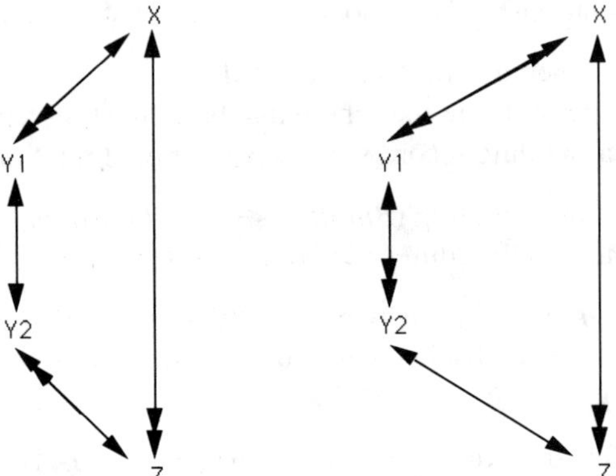

Figure 11.7 Situation where the *X–Z* relationship cannot be removed as it cannot be reconstructed from the other relationships.

There is no situation like this in our schools database diagram in Figure 11.2, but if we had a small school for younger children, the children might have only one teacher throughout the year. Then the *Teacher–Student* relation would be 1:*N* rather than *N:M* and we would have an analogous situation to Figure 11.8(*a*) in the *Teacher–Student–Course* example, as illustrated in Figure 11.8(*b*). Then either the *Teacher–Student* or the *Teacher–Course* relation could be eliminated. Probably the *Student–Course* relationship would be the one least asked in queries, so it would be the one to be eliminated. (This is different from the example in Figure 11.4(*c*) in which we removed the *N:M Teacher–Student* relation.)

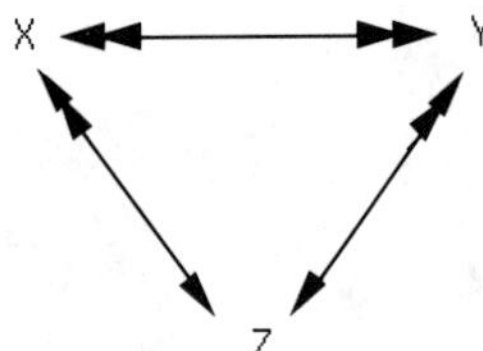

Figure 11.8(*a*) Situation where either the *X–Z* or the *Y–Z* relationship would be removed.

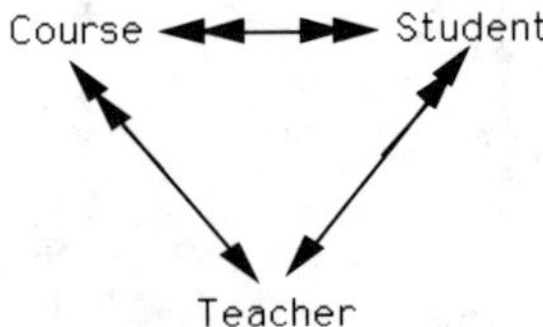

Figure 11.8(*b*) Part of a School database diagram where each student has only one teacher which is analogous to the situation in Figure 11.8(*a*). Probably the *Teacher–Course* relationship would be removed.

Note that the *N:M* relationship in this situation cannot be removed as it cannot be reconstructed from the other two. (At least one book we have read on database design states that it can.)

Rule 5. Every closed loop of relationships, in addition to those described in Figures 11.3 to 11.8, should be examined carefully to see if one of the relationships can be removed without losing information. Often one can. However, if the loop is large, made up of several relationships, then the removal of one is likely to result in substantial inefficiency. For example, in the loop in Figure 11.9 if the relationship *A–B* is removed, then any query involving *A* and *B* will require navigation through *C*, *D*, *E* and *F* which is likely to involve a lot of computing. So the relationship *A–B* might be retained on the grounds of efficiency, in spite of the redundancy created.

Returning to our school database example in Figures 11.1 and 11.2, after removing all redundant relations we obtain the much simplified entity relationship diagram in Figure 11.10.

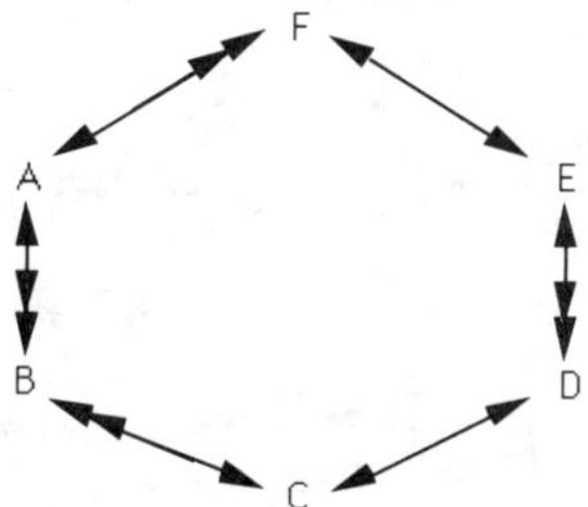

Figure 11.9 Example of a closed loop composed of several relationships. One of these *A–B* can be removed, but in such large loops this may result in substantial loss of efficiency.

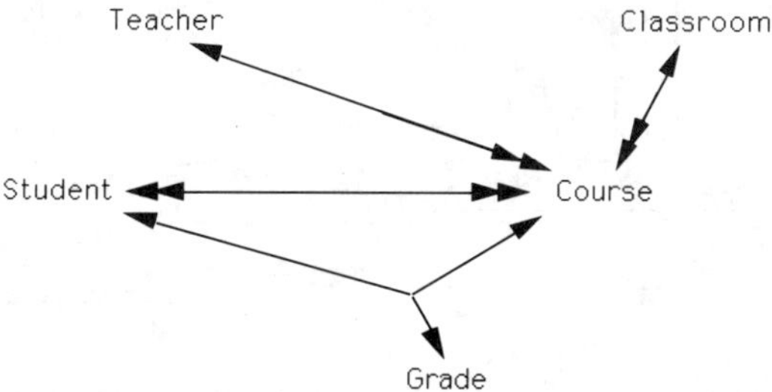

Figure 11.10 *Entity–Relationship* diagram for the school database in Figure 11.2 after the removal of unnecessary relationships

11.2.9 Realistic example

The formation of an optimal set of reduced relationships for our school example may have seemed complex on first reading. However, if you look back at section 11.2.3, you will see that it involved only six entities (including the entity Building, and about 12 relationships between them.

In practice in a database in science and particularly in engineering, you are likely to have much larger numbers of entities, numbered in 10s or 100s with even larger numbers of relationships. In Figure 11.11 we show just part of an entity relationship diagram for a database system concerned with the design of a small airplane. This shows the complexity of the problem of reducing the number of relationships in a real practical example. It is far from easy, and often the decision on which relations to remove in each loop of relationships is difficult and is to some extent subjective. It is not unknown for heated arguments to develop between the engineers taking such decisions. Yet the efficiency and effectiveness of the database system depend critically on the correct choice. One rule for a design engineer is always to seek the advice of the final users of the system when confronted with fairly evenly balanced decisions.

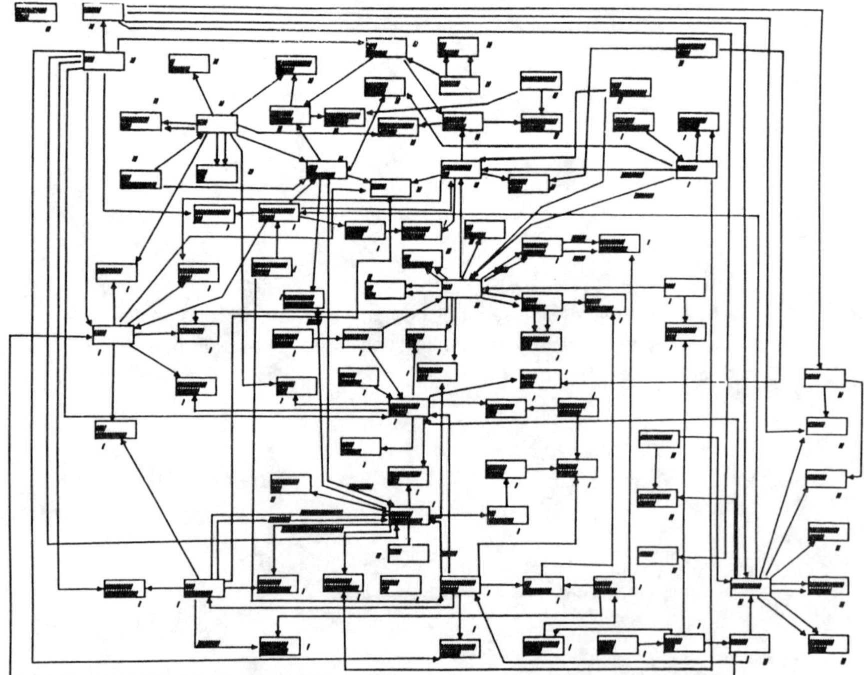

Figure 11.11 Example of an engineering entity–relationship diagram, for part of a small airplane (our thanks to Short Bros.) showing that these diagrams can become very complicated. The text has been blanked out to keep the diagram confidential.

11.3 THE RELATIONAL MODEL

11.3.1 Forming relations from the entity–relationship diagram

After removing all unnecessary relationships in the entity–relationship diagram, we can now construct in the Relational model a set of relations to represent the database. First of all one relation is formed for each entity, when it has a number of attributes. For example, in our school database the following relations for the entities *Teacher* and *Student* are formed:

1. *Teacher* (<u>*Teacher-name*</u>, *Teacher-address*)
2. *Student* (<u>*Student-name*</u>, *Grade-average*, *Age*,
                        Student-address, *Sex*).

The key field has been underlined.

Next, one relation is formed for each relationship. In the school example these become,

3. *Course–teacher* (<u>*Course*</u>, *Teacher-name*)
4. *Course–room* (<u>*Course*</u>, *Classroom*)
5. *Student–course* (<u>*Student-name*</u>, <u>*Course*</u>)
6. *Grade* (<u>*Student-name*</u>, <u>*Course*</u>, *Grade*)

Note that the key is always at the 'many' end of a one-to-many relationship and it consists of both entities together at the ends of a many-to-many relationship. For example, the key for relation 5 is the combination of *Student-name* and *Course*, since neither uniquely identifies any record in the relation.

If any two relations have the same key, uniquely identifying records in each relation, then they can be combined. For example, Relations 3 and 4 could be combined into the relation

3a. *Course–Teacher–Room* (<u>*Course*</u>, *Teacher-name*, *Classroom*).

Similarly, the relation 6, *Grade*, includes all of the information in relation 5, *Student–Course*, so relation 5 would be deleted. With experience, relation 5 would not have been considered because of relation 6.

Sometimes it is better to keep two relations with the same key separate; for example, when queries are very frequent from one relation but seldom from the second, it is more efficient to keep the frequent relation separate. Common-sense is also a factor in such decisions. It would normally be better to keep employees' medical data in a separate relation from their payroll data, even though both relations use an employee name or number as key.

After all these steps, if the entity–relationship diagram has been properly reduced to a minimum number of relationships and if database relations have been formed correctly from the relationships then the task of the initial design of the relational database is complete. However, there is an alternative procedure called normalization which can be followed to obtain the same result. In case errors have been made in the above entity–relationship process it is advisable to follow this normalization procedure once relations have been created.

11.3.2 Good relations—normalization

One of the aims of a well-structured database system is the elimination of redundancy, or rather keeping redundancy as low as possible. This aim is not in the interests of efficiency or to save disk space, but rather to eliminate inconsistencies which arise when one copy of a data item is changed and another copy of the same data item, somewhere else in the database, is not changed. For the scientist or engineer, this may result in the use of an out-of-date measurement of some physical constant or property. The new measurement may have been recorded in the database, but, because of human error, only one of the recorded values of the measurement is updated and Sod's law will ensure that you use the wrong one!

The technique used to keep redundancy as low as possible, is called normalization. Computer science textbooks on database systems[7,8,9] will describe several levels of normalization, called normal forms. Broadly these depend on two simple rules:

(i) That each field of each record contains one and only one value. (Note that this value may be a permitted word or symbol indicating that the data value is not available, or not applicable, or just null as in Table 9.2(*b*).)

(ii) That each record has one key value on which all of the other fields depend directly. (Note that since all of the fields of a record are determined once a key is known, the key uniquely defines one and only one record in the relation.)

Generally this second rule keeps relations small. For example, let us look again at the atomic physics example, the *Atomic-Mass* relation. Let us assume that we wish also to add data on the energy levels of the atoms. There are several of these for each atom so we might have chosen a relation such as the one in Table 11.1. This no longer can be given a single key. We cannot choose either the *Symbol* or *Atomic-No* as keys as the quantum number, m, and energy levels do not depend on these alone. Also we cannot choose a combination of atomic number and quantum number because the atomic weight does not depend on m.

To produce relations in normal form, i.e. 'good' relations we project out the 2 relations in Table 11.2. The key to the first can be either the atomic number or the symbol. The key to the second is the atomic number and quantum number, m, together since, if these two attributes are known, the record and the energy level are uniquely determined. If you still wish to print a table of the larger relation in Table 11.1 then you can produce this by joining the two relations together (see Section 9.4).

Note that the two new relations are smaller and simpler. It should be a general rule that you should keep relations small and simple like those in Table 11.2. However, if typical user queries require the frequent joining of relations, this may not be the best design.

Table 11.1 Example of a 'bad' relation, which does not have one unique key for each record. (m is the quantum number.)

Symbol	Atom-No	Atom-Wt	m	Energy-Level (eV)
H	1	1.008	0	0
H	1	1.008	1	10.2
H	1	1.008	2	12.1
H	1	1.008	3	12.8
He	2	4.003	0	0
He	2	4.003	1	4.5

Table 11.2 Two relations representing the data in Table 11.2. Each of these has one key, the atomic number in the first and the combination of atomic number and quantum number, m, in the second.

Relation: *Atomic Weight*			Relation: *Atomic Energy Level*		
Atom-No	Symbol	Atom-Wt	Atom-No	m	Energy Levels (eV)
1	H	1.008	1	0	0
2	He	4.003	1	1	10.2
			1	2	12.1
			1	3	12.8
			2	0	0
			2	1	4.5

Looking back at the School database in Section 11.2, the final relations

1. *Teacher (<u>Teacher-name</u>, Teacher-address)*
2. *Student (<u>Student-name</u>, Grade-average, Age, Student-address, Sex)*
3a. *Course–Teacher–Room (<u>Course</u>, Teacher-name, Classroom)*
6. *Grade (<u>Student-name</u>, <u>Course</u>, Grade)*

all have one key which uniquely defines all of the other fields in each relation; so these are already in normal form.

11.4 HIERARCHICAL MODEL

In the Hierarchical model (see Section 9.2) all data are structured in trees, i.e. in series of 1:N relationships *Father–Son, Son–Grandson*, and so on.

The entities *Father, Son*, etc., are represented by records, which may have two or more fields. Only one field (usually the entity name or number), is used when the entity has no other attributes.

The hierarchical model does not support *N:M* relationships (without massive redundancy), therefore as described in Section 9.2 any *N:M* relationships are replaced by two 1:*N* relationships, which then become two separate trees or parts of trees. For example, in our school database in Figure 11.10 the *N:M Student–Course* relationship is replaced by the 1:*N* relationships:

> *Course ⟷≫ Course–Student–Grade*
> *Student ⟷≫ Course–Student–Grade.*

To find the trees representing the structures in the entity–relationship diagram, we look at each 1:*N* relationship. The higher or *Father* entity in the relationship is examined to see if it has any *Grandfather* entities defined by either a 1:*N* or 1:1 relationship. If it has only one *Grandfather*, then the *Grandfather–father–son* entities and relationships form a tree or part of a tree in the database. The same process can be continued higher or lower levels and a tree of several levels can be formed.

When any entity is a child to two or more different entities then two separate trees must be formed. If the child entity has lower grandchild entities, then to avoid redundancy, these should be attached to only one of the two trees, the one which is most likely to answer user queries effectively and efficiently.

In the school database described in the entity–relationship diagram in Figure 11.10, the Hierarchical structure could be the one shown in Figure 11.12. The top of the main tree is chosen to be *Teacher* rather than *Classroom* because more queries are likely about *Teacher*, *Courses* and *Grades* than about *Classroom*, *Courses* and *Grades*.

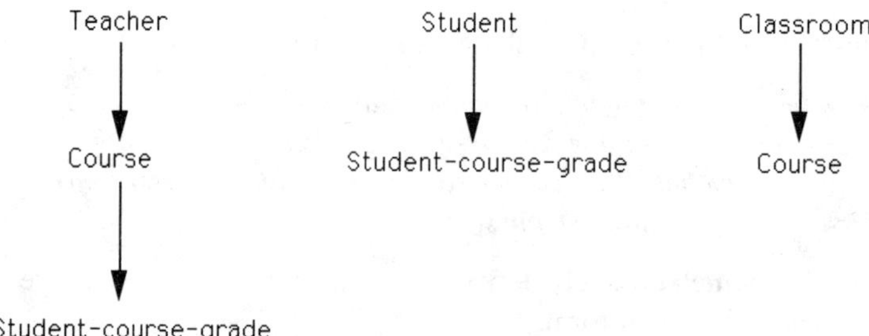

Figure 11.12 Hierarchical model of our school database in Figure 11.10.

11.5 NETWORK MODEL

In the Network model (see Section 4.3) all data are represented in the form of sets consisting of trees with two levels, owner records and lower member records, corresponding to 1:*N* or 1:1 relationships.

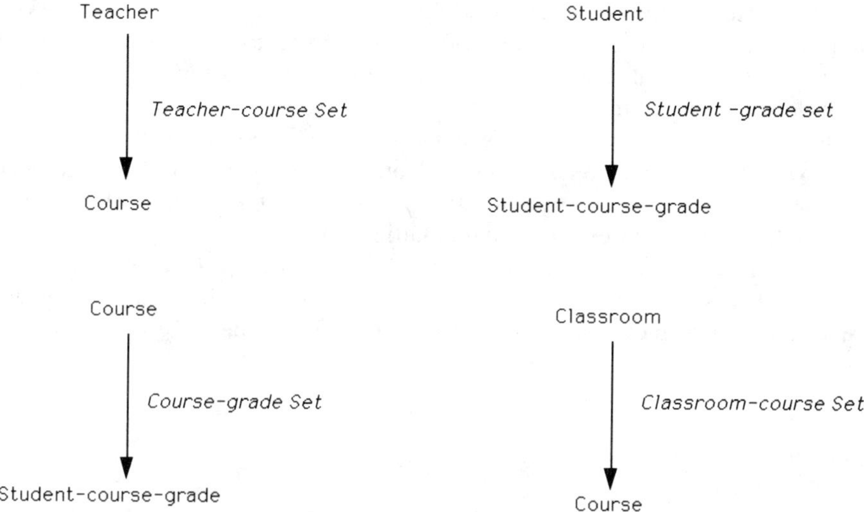

Figure 11.13 Network model of the school database in Figure 11.10 and 11.12.

As in the hierarchical model, entities are represented by records with one or more fields. The Network model does not support $N{:}M$ relationships and these must first be reduced to two $1{:}N$ relationships by the introduction of link records (see Section 4.3.2). Then each $1{:}N$ or $1{:}1$ relationship becomes a set. So the process is slightly easier than the creation of a hierarchical database.

The school database in Figure 11.10 is represented by the sets in Figure 11.13.

11.6   INVERTED LIST MODEL

In the Inverted List model all data are stored in files, similar to relations. The procedure followed for the identification of relations described in Section 11.3 is followed, each relation becoming a file in this model.

Therefore, for the school database, the inverted list files would be equivalent to the four relations at the end of Section 11.3, i.e. files corresponding to the *Teacher, Student, Course–Teacher–Room* and *Grade* relations.

11.7   PROTOTYPING

The design described in the previous sections depends on an accurate specification of the user needs and priorities. In practice, it is extremely difficult to discover exactly what are the user needs because they are usually not at all clear to the users themselves and because communication between the users and designer cannot be perfect. It is therefore advisable to build a prototype system at an early stage, before large amounts of data are added to the database. The prototype system would use sample files of real data and should give some users the opportunity to use at least a cut-down version of the final system.

Usually, when users see examples of questions and responses obtained from the prototype system on the screen or in printed form, it is only then that it becomes clear that their needs are not those initially specified, and what the needs really are; the system then often requires substantial modification.

When changes are made, they should be made from the top down, redrawing the entity–relationship diagram, removing unnecessary relationships (not always the same removed as previously), normalizing, breaking some large relations with many fields into two or more relations, and so on.

11.8   SIMULATION

It is important to obtain some idea of the performance of the system when it becomes finally in full operation, in addition to user response to the prototype. It is normally not possible to begin tests with a complete database because the data are not available until too late in the design process. However, it is usually not difficult to simulate the performance of the final system. One method which is straightforward, is to write software to generate large numbers of random records to fill the relations or sets to the expected size. Alternatively a small file can be appended to itself and thus doubled in size. After doubling 10 times it is 1000 times larger than it was at first. A few additional records can be added for tests and measurements made of times of searches, joins of relations, navigation through trees, or sets all performed to find how the system will react to typical user queries.

Many a database system has been sold because of its response to a search in a file of 100 records—but it was a disaster for the 10 000 records needed. We know of one example where a substantial investment was based on a system which in order to answer user queries depended on a join of two relations, and it was not realized that the join with the final files (20 000 records long) would be slower than the test files displayed by the salesman.

Simulation of the final full system can often be undertaken in a matter of days. It is always worthwhile; and the results can lead to major changes in the design. Some redundancy may have to be accepted and some relationships retained which were removed from the earlier entity–relationship diagram, either because of user need or increased efficiency.

11.9   DATA DICTIONARY AND THESAURUS

The data dictionary (Section 8.3) and thesaurus (Section 8.4) are important user facilities and need to be built into the design, at an early stage. If your database system does not provide these facilities, then a separate database should be set up in parallel with the main database recording the metadata of the data dictionary. This is worth doing even for the smallest database. At the very least these data should be recorded on paper, in a manual or other booklet. But this is likely to get separated from the database, so it is better for it to be stored with the database.

A thesaurus is less important than the data dictionary in small databases and is not needed in many systems. For large online systems which integrate several databases together, a thesaurus may be indispensable.

11.10   MEASUREMENT AND FEEDBACK

Finally, the design should include facilities to enable the performance of the database to be measured regularly and for adjustments to be made when inefficiencies develop or user queries change from the queries originally specified.

It is also important to find out not only how the computer system is performing as seen by those in charge of the system, but also how the users perceive it to be performing. User views of the system and changing user needs are probably the most important factors (although unfortunately the least often measured factors) which should provide regular feedback for modifications to the design at all levels at regular intervals, say once a year.

REFERENCES

[1]   Accreditation Board for Engineering and Technology, 1985 Annual Report, 345 East 47th St, NY, 4–115, 1985.

[2]   P. Chen, The Entity–Relationship Model—towards a Unified View of Data, *ACM Trans. Database Systems*, 1, 9–36, 1976.

[3]   P. Chen, *The Entity–Relationship Approach to Logical Database Design*, QED Information Sciences, Wellesy, Mass, 1972.

[4]   H. Robinson, *Database Analysis and Design*, Chartwell-Brett, Old-Orchard, Kent, England, Chapter 7, 1981.

[5]   P. Yourdon, *Modern Structured Analysis*, Prentice-Hall, Englewood Cliffs, NJ, 1989.

[6]   R. Elmasri and S. B. Novathe, *Fundamentals of Database Systems*, Benjamin/Cummings Publ. Co., Redwood City, Ca, Chapter 3, 1989.

[7]   R. Elmasri and S. B. Novathe, *Fundamentals of Database Systems*, Benjamin/Cummings Publ. Co., Redwood City, Ca, Chapter 13, 1989.

[8]   H. Robinson, *Database Analysis and Design*, Chartwell-Brett, Old-Orchard, Kent, England, Section 7.4, Chapter 7, 1981.

[9]   B. J. Salzberg, *An Introduction to Database Design*, Academic Press, Orlando, Florida, Chapter 2, 1986.

Chapter 12

Expert Systems and Knowledge Base Systems

12.1 INTRODUCTION

A database system is a system for the storage and retrieval of data. A user may search the database for data to solve some problem, but the database system does not solve the problem. All it can do is retrieve data for the user and possibly manipulate the data (such as change units) to help the user solve the problem. On the other hand, an expert system stores rules (usually logical rules) as well as data in the form of 'facts' but its purpose is not to provide a system to retrieve the rules and facts, a relatively easy task, but rather to use the rules and facts together to find a solution or solutions to the problems posed by the users.

Normally the rules are based on the experience of an expert, and the system attempts to emulate the expert's ability to solve problems—hence the term expert system. Because the task of an expert system is so much more complex than that of a database system, it normally operates in a precise and very limited domain, i.e. on a knowledge base of only a few hundred facts and rules, whereas a database system has no difficulty in dealing with tens of thousands of records.

Sometimes the knowledge base consists of a large amount of data and a small number of rules. Then a combination of database and expert system may be needed. This is discussed at the end of this chapter, but first we discuss the common form of expert system which infers conclusions from logical rules; this is the form of expert system which can be purchased easily for either small or large computers. The discussion must be brief, as this book is primarily about database systems; but we think that it is important to include an overview of the subject here because there is an overlap between the two and because you may find at the planning stage of your information system that an expert system or a mixture of expert and database systems is what you need rather than a simple database system.

Since this chapter must be brief, we refer the reader to other texts for a wider understanding of the subject. There are now many books on expert systems or knowledge based systems on the market, although many of these are essentially reports on conference proceedings which report on research on the subject and are not easy to read and evaluate by the inexperienced reader. There are, however, a small number of shorter texts which give good introductions to the subject. One which is simple and straightforward and very readable is entitled *Build your own Expert System* by Chris Naylor[1]. It is also particularly suited to the scientist or engineer as it explains many of its concepts in terms of simple mathematics which should be easily followed by a scientist. Another good text covering much of the same ground and including details of several working systems is by Alty and Coombs[2] and a further book, with a broader view of knowledge based systems by Black[3] is also recommended. Many books on artificial intelligence contain relevant chapters on expert systems: one worth reading is in *Intelligent Knowledge-Based Systems* edited by O'Shea, Self and Thomas[4].

12.2 PRODUCTION RULES

Expert systems are usually based on a set of 'rules of thumb' called production rules which represent the accumulated experience of an expert in a particular subject. We illustrate with an example. Most of us use PCs and experience has taught us the cause of most problems which occur—so we become 'experts' on its use. We look at the frustrating problem when we send a file to the printer and nothing happens. Experience tells us that:

   (i) the printer is not switched on,
  (ii) the ready button on the printer has not been pressed,
 (iii) the printer is out of paper, or
 (iv) the connection at the back of the printer is loose.

This simple example illustrates one characteristic of many rules, that they involve an amount of uncertainty, as we know that these causes have different likelihoods or probabilities. Your experience with your PC can be formalized into several rules including the following:

   (i) if the printer does not print then there is a high probability 0.7 that it is not switched on

  (ii) if the printer does not print then there is a small probability 0.1 that the printer is out of paper

 (iii) if the ready button has not been pressed, the printer will not print

and so on.

The first two rules take the form

     if <antecedent> then <consequence> has probability p

or   if <premise fact F> then <hypothesis fact C>, probability p

or some similar terminology. In the second expression above <premise fact F> reads 'premise fact F is true' so that the whole line reads:

if premise fact F is true then the hypothesis that fact C is true has probability p

This can be abbreviated in different ways. One convenient abbreviation is

F $\rightarrow$ C   probability p

So returning to the three rules numbered, the first two have the above form, but the third has the simpler form

if F is true then C is true
or   F $\rightarrow$ C

where F is the fact (or fault) that the ready button has not been pressed. In this case the probability is unity and need not be included.

It is common in expert systems to use a measure of probability called 'certainty', partly because most people not trained in mathematics find it difficult to understand the concept of probability. Certainty is measured with a scale from -1 to +1 or from -5 to +5 or some similar scale. This is a linear scaling of probability such that a probability of 0.5, when an event is neither likely nor unlikely, has a certainty measure of zero. Positive certainties are equivalent to probabilities greater than 0.5 (for an event which is likely) and negative certainties are equivalent to probabilities less than 0.5 (for an event which is not likely). This is illustrated in Figure 12.1.

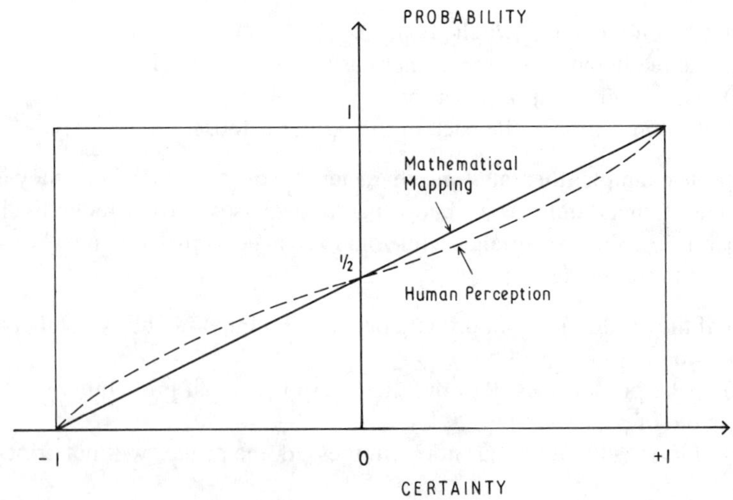

Figure 12.1 Illustration of how a mathematical certainty measure from -1 to +1 can be mapped onto probability in (0, 1). However, a human perception of certainty does not always accord exactly with this mathematical definition.

As we have begun to show, an expert system operates on a set of rules acting on a number of facts. In our simple example, which illustrates a much simplified version of a fault-finding expert system, we have facts which we define as follows:

C: the printer does not print correctly,
R: the ready light is off,
F_1: the printer is not switched on,
F_2: the ready button has not been pressed,
F_3: the printer is out of paper,
F_4: the connection is loose,
F_5: some system fault.

To be more precise, these facts C, R, F_1, etc. are assertions that the particular fact is true and they therefore have the value either True or False. Along with the facts are rules concerning these facts. For example, the three rules already mentioned can be written

$$C \rightarrow F_1 \quad \text{probability } 0.7$$
$$C \rightarrow F_3 \quad \text{probability } 0.1$$
$$R \rightarrow C$$

Three other rules state that certain faults result in the ready light being off (R):

$$F_1 \rightarrow R$$
$$F_2 \rightarrow R$$
$$F_3 \rightarrow R$$

and a subsequent rule states that the printer will not print if the ready light is off:

$$R \rightarrow C.$$

Further rules exist for the remaining faults, F_4 and F_5:

$$F_4 \rightarrow C \quad \text{probability } 0.7$$
$$F_5 \rightarrow C.$$

The first of these three has a probability 0.7 since a loose connection may not always cause a fault.

There are also inverse rules to the above, for example, the inverse of the last rule, in a more useful form, tells us that when the printer does not work it is very rarely a system fault:

$$C \rightarrow F_5 \quad \text{probability } 0.0001$$

Such a collection of rules could represent our experience of faults on the printer. A very small expert system could be written to incorporate your experience. It would do this by use of a special computer program written to simulate this experience. Because the system would infer the nature of the fault from information given, the software system which operates on the rules is called an inference engine.

12.3 INFERENCE ENGINE

We can summarize the principles in the last section by the expression

software + facts + rules = expert system.

The rules and software are kept separate from one another in the same way as the data and programs are kept separate from one another in a database system. The rules which constitute part of the expert system are called static rules as they do not change, even after the expert system is used to solve a number of problems.

To solve a problem some additional facts (and possibly rules) have to be added to describe the problem. They are called variable data as they vary from problem to problem. In a medical environment these would be the symptoms. In our simple printer example, they would be facts about the printer such as

C is true
(the printer does not work)
and
R is true
(the ready light is not on).

From these variable facts the expert system would infer the most likely faults and it may pose a series of questions to obtain more data such as

Is F_3 false?
(Is there paper in the printer?)

it will eventually narrow down the likely fault and thus advise on the correct action, e.g. press the 'ready' button,

This process is demonstrated in Figure 12.2.

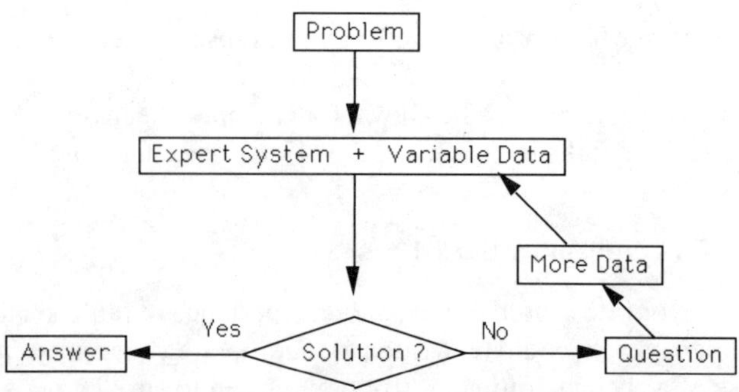

Figure 12.2 This illustrates how variable data (symptoms) together with an expert system invoke a series of questions followed by an inferred result.

12.4 CHAINING

There are two main approaches followed by the software of an inference engine:
backward chaining and forward chaining. These are discussed separately below.

12.4.1 Backward chaining

To illustrate the process of backward chaining we first look at our simple printer
example. We want to find out the cause of the printer failure (G true) so we begin
with this goal and start by looking for rules which infer G. The first of these is

$$R \rightarrow G$$
(if the ready light is off, the printer will fail).

Now R is unknown so the system first asks if R is true, using an English language
sentence:

'Is the ready light off?'

If the answer is yes, then R is true and the system next looks at rules which infer R
is true. There are three, the first of which is

$$F_1 \rightarrow R$$
(if the ready button is not pressed, the ready light is off)

So the system asks if F_1 is false (has the ready button been pressed?) If the answer
is no; then the system advises the action 'Press the ready button'. The program has
completed its task and given advice on action to take.

The above example is very simple, with only one backward link in the chain; so
we illustrate backward chaining with another slightly more complex example concerning seven facts—A, B, C, D, E, F, G and the following six rules.

1. A and B $\rightarrow$ C
2. C and D $\rightarrow$ E
3. F $\rightarrow$ E
4. F and B $\rightarrow$ G
5. A and E $\rightarrow$ H
6. B and E $\rightarrow$ G

The first rule reads:

if A is true and B is true then C is true.

Of course the logical process we are about to describe does not depend in any way
on the nature of the facts A, B, C, etc or of the six rules. For example, the first rule
might mean that if a bird is brown (A) and with a red breast (B), then the bird is a
robin (C). Alternatively it might mean that if your nose is running (A) and you have
a headache (B), then you have a cold (C). Adding in the meaning of the rules

obscures the logical process followed by the computer, which always views the set of facts and list of rules in their abstract form, that is, a set of Boolean variables A to G and as a list of logical rules. The 'meaning' of the facts is only stored in the computer in the form of a string of characters for each fact, a string which has meaning to a human reading it, e.g. B: 'the bird has a red breast'.

Now let us assume that we add to the knowledge base of six rules above, two variable facts:

$$B \text{ is true}$$
$$D \text{ is true}$$

and we want to know if G is true. The inference engine begins with G and first looks at which rules would infer G is true. These are rules 4 and 6. So it starts with the first of these, rule 4 and noting that B is true finds that rule 4

$$F \text{ and } B \rightarrow G.$$

becomes the new temporary rule:

$$T_1 \quad F \rightarrow G$$

It notes that F is unknown, and that there are no rules inferring F (i.e. F is not on the Right Hand Side of any of the above rules). Since it cannot move backwards any further and F is unknown it must try another path. So it next tries rule 6 which might also infer the truth of G:

$$6. \quad B \text{ and } E \rightarrow G$$

This becomes:

$$T_2 \quad E \rightarrow G \qquad \text{since B is true}$$

and chaining backwards using rule 2, we derive a third temporary rule:

$$T_3 \quad C \text{ and } D \rightarrow G \qquad \text{using rule 2.}$$

This is replaced by

$$T_3 \quad C \rightarrow G \qquad \text{since D is true,}$$

and since C is unknown it moves backwards another step to find the new rule

$$T_4 \quad A \text{ and } B \rightarrow G \qquad \text{using rule 1.}$$

This is replaced by

$$T_4 \quad A \rightarrow G \qquad \text{since B is true.}$$

Since A is unknown and no other rules infer A, this is as far back as this second chain will reach.

By this process of backward chaining the inference engine has thus inferred four new rules T_1 to T_4. However, since F, E, C and A are unknown the system still cannot determine G without more data. So it asks

Is F true?

Of course the question is not put in this form. Every expert system has stored along with its facts and rules, English expressions for all of the possible questions which the system can ask or a simple language generator based on words, phrases and a simple syntax which can generate the English questions. These are generated or stored in the knowledge base to improve the user interface (and to make the system look more intelligent than it is). For example, the above question might be asked in the form

Is the paper loaded correctly in the printer?

or

Do you have a headache?

If the answer is yes, then the truth of G is inferred from rule T_1. If the answer is no, i.e. that F is false, then rule T_1 does not infer its converse, that G is false. For example, if F is that the printer is switched off, and G that the printer does not work then clearly

$F \rightarrow G$

(if the printer is switched off, it doesn't work)

But if F is false (switched on), G may not be false, the printer may still not work). So

F false does not imply G false

So an answer of 'no' or 'don't know' to the question 'Is F true?' forces the system to go to the next rule T2 and ask

Is E true?

In a similar manner it may go on to rules T_3 and T_4 asking if C or A are true. If E, C or A are known to be true, G is inferred to be true. Otherwise no decision can be taken.

This process of backward chaining, starting with a goal and searching back through the rules and facts, and interacting with the user to prove or otherwise that the goal is true, is one of the most commonly used processes in expert systems.

As we shall see it becomes more complicated when probability is added.

12.4.2 Forward chaining

The reverse process, forward chaining, is less common. In this the system begins with all the known variable facts (symptoms) (in our example: B true and D true), and by forward chaining from rules which include either B or D as antecedents deduces new facts and rules. This process continues along these chains which fan out in a tree until the truth of the goal is established or until the process is exhausted. In our example we begin with rules 1 and 2 giving the following two new rules S_1 and S_2:

         1.   A and B   →   C

becomes

         S_1 A   →   C   because B is true

and

         2.   C and D   →   E

becomes

         S_2 C   →   E   because D is true

Using rules S_1 and S_2 together we can deduce a new rule:

         S_3 A   →   E

Similarly other new rules can be deduced from rules 3, 4, 5 and 6 and these will include the rules we obtained by backward chaining. However, the process is less efficient than backward chaining because many unnecessary rules will be generated, not relevant to the truth of the goal G. In a large knowledge base, very large numbers of additional rules are generated, taking considerable processing time, and these are not easy to store (or index). Sometimes this process can generate new rules, not previously anticipated by experts, which may be useful and enlightening to the experts. However, for problem solving, the main aim of an expert system, backward chaining is faster and is generally preferred.

12.5   EXPLANATION

To improve the man–machine interface, facilities are normally included in the expert system to help explain to the user what the computer is doing. Two facilities are commonly included, based on the questions Why? and How?

Why?

In the backward chaining example above the user might ask 'Why?' in response to the question:

         Expert System   Is A true?
         User   Why?

Then the system would explain giving all facts and rules in the backward chain leading to the question:

         Because if A is true since B also is true,
         so by rule 1

                  A and B   →   C

         so C would be true.

         But D also is true, so by rule 2

                  C and D   →   E

so E would be true.

But B also is true, so by rule 6

> B and E → G

so G would be true, which would answer your question.

> Is A true? (Asking the question again.)

Of course, all of these statements would be written out in the form of English sentences or statements which would be stored within the expert system or generated with a simple syntax.

How?

If the user is given a final conclusion, e.g.

> G is true

then the user might ask 'How?'. Then the system should give all the previous rules and facts in the backward chain used to obtain its conclusion. The response would be very similar to the response given above in the example on a Why? query, so we need not list them again.

These two explanations based on Why? and How? requests are generally the minimum facility of this kind in an expert system.

12.6 PROBABILITY (OR CERTAINTY)

In practice probability is nearly always an essential part of the rules (and sometimes of the facts) within a knowledge base. As we explained in section 12.2 and in Figure 12.1, expert systems normally express probability in terms of certainty, which is more easily understood by people not familiar with the concept of probability. In the following we must use probability because it is mathematically precise.

In its most general form a rule can be written:

> If <f has probability p_f> then <c has probability p_c>

where p_c depends on p_f. We examine now a few of the problems and discuss some possible solutions.

Multiplication rule

Unfortunately when we know p_f, it is rare that we know an exact mathematical formula which enables us to calculate p_c. However, often it is a good approximation to use a simple multiplication rule. If the original rule takes the form

> If <f> then <c has probability p_c>

and if f is not known with certainty but has a probability p_f then we can assume to a good approximation that the new probability of c, p_c' is the product of p_f and p_c, i.e.

$$p_c' = p_f p_c.$$

Let us illustrate with a simplified example. We assume that it is known that an acute pain in a certain part of the lower abdomen (fact f) indicates the consequence (c) that a patient has an appendicitis, with a high probability, $p_c = 0.8$. However, a doctor may find the pain is not quite in the correct place and appears to be less severe than would be expected. In this case there would only be a probability, say 0.7, that this is the correct symptom ($p_f = 0.7$). The probability that the patient has an appendicitis (p_c) would then be given approximately by

$$p_c' = p_f p_c = 0.8 \times 0.7 = 0.56$$

The doctor would now look for other symptoms (i.e. other facts) before sharpening the knife!

Compound antecedents

Often the antecedent of a rule involves several facts, not just one. Let us first look at the case of two facts h and g and discuss how their probabilities can be combined.

h and g

Let us assume that there are two facts h and g pointing to a consequence c, i.e. we assume f is made up of h and g together:

$$f = h \text{ and } g.$$

For example h and g may be two different salts of some metal and c the fact that there are mineral bearing seams in the neighbourhood. We also assume that they have probabilities p_h and p_g respectively. The first, p_h, in our example is the probability that the presence of salt, h, indicates a mineral bearing seam. Similarly for p_g.

If h and g have no correlation (independent of one another), then the probability of the two together equals the product of the probabilities.

$$p_f = p_h p_g$$

and if there is perfect correlation (one fact always accompanied with the other) then

$$p_f = p_h = p_g.$$

But in practice there is usually some correlation between the facts (such as pain in the abdomen and nausea), but the amount of correlation is unknown. In this case limits can be set on the probability, set by the extremes of no correlation for a lower bound and perfect correlation for an upper bound. For the upper bound, since ideally

p_h should equal p_g, we set the limit to be the lower of the two. So

$$p_h p_g \leq p_f \leq \min(p_h, p_g) \tag{12.3}$$

The higher value

$$p_f = \min(p_h, p_g) \tag{12.4}$$

is used effectively as an approximation to p_f in several expert systems such as the geology expert system, Prospector[9].

h or g

The antecedent f may be made up of one fact h **or** another fact g; for example h or g may mean that either salt h or salt g is found in the soil. Then either may indicate a mineral resource.

If the two facts are independent then the probability of

$$f = h \text{ or } g$$

equals the probability that both **h** and **g** are not false, i.e. it equals unity (i.e. certainty) less the probability that both facts are false. Now the probability that h is false is given by

$$p_h \text{ false} = (1 - p_h),$$

similarly for g; so

$$p_f = 1 - (1 - p_h)(1 - p_g)$$
$$= p_h + p_g - p_h p_g. \tag{12.5}$$

This sets the upper limit when the correlation is unknown. If there is perfect correlation then $p_h = p_g$. But for an unknown correlation, we note:

$$p(h \text{ or } g) \leq \text{ either } p_h \text{ or } p_g,$$
So
$$p(h \text{ or } g) \leq \max(p_h, p_g).$$

This sets the lower limit; so we conclude:

$$\max(p_h, p_g) \leq p_f \leq p_h + p_g - p_h p_g \tag{12.6}$$

The lower limit

$$p_f = \max(p_h, p_g) \tag{12.7}$$

is often used as an approximation in expert systems, for example in Prospector.

It is not difficult to find limits and approximations to the probability of a combination of three or more facts by an extension of the above mathematical analysis. However, the uncertainty and accuracy of the composite probabilities obtained become more open to question as the number increases.

The above discussion will illustrate the difficulties of using probabilities in expert systems and show that they cannot give exact or even nearly exact results; there is greater uncertainty than you find in most branches of science and engineering.

12.7 BAYES' THEOREM

Up to now we have dealt with a rule of the form

If <f> then <c has probability p_c>

and introduced a modification if f has probability p_f rather than unity. We have also looked at the problem of combined facts in one composite antecedent f. In these discussions we have assumed that initially p_c is unknown. However, we need to consider the case when c has a prior probability p_c before introducing fact f. For example, if c is the disease measles, then generally the probability that a child entering a doctor's office has measles is very low. So $p_c \approx 0$. However, if there is a measles epidemic in the neighbourhood, p_c will be a lot higher (say $p_c = 0.3$) and this will affect the final probability p_c' after the child is seen to have a spotty face. So we need to find a means of including the prior probability p_c. Bayes' theorem enables us to do this. It has another advantage. Probability or even certainty are not ideas which are easy for some experts to understand and estimate. However, the concept of likelihood, finding the ratio of two possibilities, is much easier to understand as it is similar to the idea of betting odds. Bayes' theorem enables us to introduce this concept.

Let us assume that some prior fact h (a measles epidemic) has the consequence c with probability p_c, i.e.

$$h \rightarrow c \text{ probability } p_c$$

This can be written mathematically as

$$p(c \mid h) = p_c$$

where the expression $p(c \mid h)$ means the probability that fact c is true if fact h is true.

Now introduce another independent fact f (a spotty face) with probability p_f and as we have stated, before it is introduced c has the prior (*a priori*) probability p_c. The probability that c and f are both true can be written

$$p(c \text{ and } f) = p_f \, p(c \mid f) \tag{12.8}$$

i.e. it equals the probability of f being true times the probability that c is true if f is true.

Similarly

$$p(c \text{ and } f) = p_c \, p(f \mid c) \tag{12.9}$$

Equating (12.8) and (12.9) we get

$$p_c \, p(f \mid c) = p_f \, p(c \mid f)$$

or

$$p(c \mid f) = p_c \, p(f \mid c)/p_f \tag{12.10}$$

This is Bayes' theorem. If the denominator is written:

$$p_f = p_c\, p(f\,|\,c) + p_{not\text{-}c}\, p(f\,|\,not\text{-}c) \qquad (12.11)$$

it allows us to introduce the new likelihood ratio:

$$LS = \frac{p(f\,|\,c)}{p(f\,|\,not\text{-}c)} \qquad (12.12)$$

This is the likelihood that the fact f (some symptom, like spots) is due to consequence c (some disease, like measles) compared to not-c (not the disease, not measles). As we have mentioned earlier these likelihood ratios can be estimated more easily by most people, including an expert than probabilities. We can write $p(c\,|\,f)$ in terms of LS using (12.10), (12.11) and (12.12), giving a new form to Bayes' theorem:

$$p(c\,|\,f) = \frac{LSp_c}{LSp_c + 1 - p_c} \qquad (12.13)$$

$$p\,(c\ and\ f) = p_f\, p(c\,|\,f)$$

$$= \frac{LS\, p_c p_f}{1 + LS\, p_c - p_c} \qquad (12.14)$$

Other similar formulae can be derived for the introduction of facts f_2, f_3, etc.

A complete analysis involves the probability of consequence c if f is false, i.e. on the probability

$$p(c\,|\,not\text{-}f)$$

and on the likelihood ratio that not-f is caused by c rather than by not-c:

$$LN = \frac{p(not\text{-}f\,|\,c)}{p(not\text{-}f\,|\,not\text{-}c)} \qquad (12.10)$$

For example, in a medical expert system, this is the probability you do not have a symptom with a particular disease (no spots if you have measles) compared to the general probability of not having the symptom (no spots if you do not have measles). The ratio LN is estimated by the expert and stored with the other likelihood value LS for each antecedent fact f and consequence c. It is usually very small.

From the value of LS the new probability $p(c\,|\,f)$, that is the probability that the consequence, or disease c, with prior probability p_c is true if symptom f is true is calculated from equation (12.13). The probability when there is no symptom $p(c\,|\,not\text{-}f)$ is similarly computed from LN. More details are given in references [2] and [5].

The above incomplete derivation assumes that h and f are independent (no correlation) which is unlikely to be correct; nevertheless it is normally a good approximation.

We have covered only a part of the subject to illustrate the techniques typically used in expert systems. When you recall that the above incomplete derivation

(i) involves only two facts h and f

(ii) assumes that h and f are independent (no correlation) which is unlikely to be correct

at first it appears that the subject is complicated. But the mathematics are not difficult after some experience and in an expert system the mathematics are hidden from the users, although the computed probabilities obtained from backward or forward chaining are far from exact, they are still useful estimates in practice.

12.8   NUMBERS OF COMBINATIONS OF RULES

One apparent solution to the problem of lack of accuracy in probabilities obtained from an expert system is to ask an expert to estimate the probability of all suitable combinations of facts, f, and consequences c. However, it is easily shown that this is impractical because there are usually a large number, n, of different facts and the number of possible combinations of facts is excessively high as illustrated in Table 12.1. The total number of combinations for all sizes of groups together can be shown to equal 2^n, which equals about a million when $n = 20$. So, even for this small number, there is no hope of obtaining estimates of probabilities for all possible combinations of events from a human expert. For 400 facts (a typical size of an expert system), the total number of ways of combining them would be greater than 10^{120}!

12.9   TESTING

We are therefore left with empirical rules such as the ones in Sections 12.5 and 12.6.

Table 12.1 Numbers of combinations of n facts in groups showing that the numbers are very large.

Number of facts grouped together	Number of combinations
1	n
2	$\dfrac{n(n-1)}{2!}$
3	$\dfrac{n(n-1)(n-2)}{3!}$
4	$\dfrac{n(n-1)(n-2)(n-3)}{4!}$

Sometimes it is necessary to assume that the facts are independent, as an approximation, even though it is known that they are not. It is not unknown that in tests of an expert system some probabilities are found to be greater than unity!! We would not wish to recommend such an expert system; however, it is normal practice to conduct a large number of experiments on a system and allow the expert to comment on the results. When results are clearly wrong the probabilities and sometimes the mathematical rules and simplifications have to be adjusted and tuned so that the system performs in a manner similar to that expected by the expert. However, just as it is not possible to test every path through a computer program, so it is not possible to test every type of variable data presented to an expert system by a potential user. Only by extensive tests, changes and additions can the system begin to iterate towards the expertise of the human expert.

12.10   THE EXPERT

It should be abundantly clear already that the advice of the expert or experts providing information on probabilities and rules is quite critical to the success of an expert system. A group of computer scientists, no matter how brilliant or experienced, have no hope of building on their own a useful expert system on any subject, apart from an expert system on computer science itself.

The ideal team to produce a successful expert system is one consisting of specialists in the subject of the system, together with computer scientists, both highly motivated. For a small system a group of specialists might be reasonably successful on their own by using one of the off-the-shelf expert software systems (shells) available on most computer systems. However, whether the expert system be large or small, an essential ingredient for success is the motivation of at least one expert. Indeed two highly motivated specialists may be less successful than one, because two may not agree on the rules to be entered!

The motivation must be high enough that the specialist gives a great deal of time to the system. If the specialist is highly trained and his skills are much in demand (and why build the expert system if they are not?) then finding time to build up all of the necessary facts and rules will be extremely difficult.

There is one factor which makes the problem more difficult and causes many expert systems to fail (apart from the natural reluctance of any specialist to give up his expertise to a computer system which might outperform and replace him.) This is the confidence crisis which develops soon after an expert system project is started. This confidence crisis begins with an underestimate of the complexity of the system to be modelled by the expert system. The number of rules and possibly the number of facts are likely to be underestimated by a large margin—possibly by a factor of 2 or 3 or more.

Initially as the first rules are added and the first tests show the considerable potential of a rule based system, confidence and expectations grow. Then the complexity of the real expertise of the specialist begins to become apparent as more and more realistic problems are attempted and fail; so confidence drops along with enthusiasm and motivation. This is illustrated in Figure 12.3, which shows that confidence (like cash flow in a new company) drops to a minimum; it then becomes extremely difficult to convince a specialist that he should continue to give his valuable time to the project (just as it is hard to find cash at the equivalent point of a new company). Many expert systems stop there, with a performance well short of what can be attained, sufficient to give interesting demonstrations to visitors, but not sufficient to be useful in real problem solving. Success would have been higher if the objectives had been much reduced at the outset (see Section 12.14).

If the crisis can be overcome, and if the organization employing the specialist agrees to release him for sufficiently long periods of time, then a process of tutorial interaction with the specialist, testing the system with numerous examples, and studying more and more special cases will eventually refine the system. The performance will continue to improve and confidence will return.

It is unlikely that the expert system can outperform the specialist when nothing else but the experience of a specialist is being modelled. When the expert system has a wider role, including scientific calculations or retrieval from a large database, then it is quite likely eventually to outperform the specialist, in the sense that it will solve many problems faster. This is more likely in science and engineering than in other subject areas.

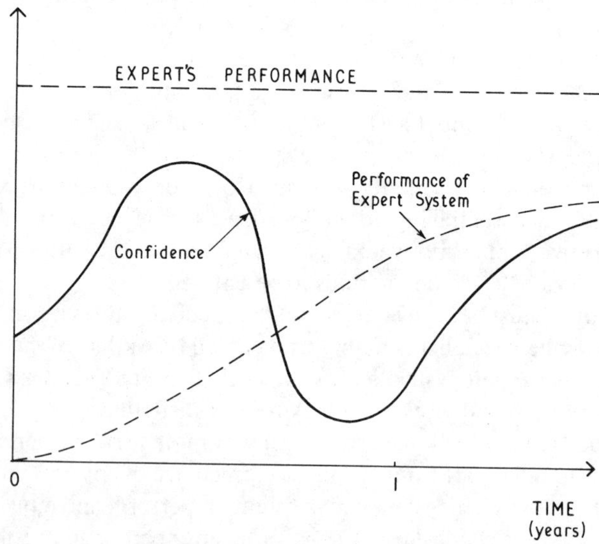

Figure 12.3 Illustration of the confidence crisis which occurs 6 months to 1 year after an expert system is started.

12.11 SUPPORT FOR THE EXPERT

It is important that the time of the specialist be used as effectively as possible. Only the expertise of a top specialist will be worth modelling, but a top specialist's time will be very difficult to obtain. Therefore it has been found essential that special software be provided to enable the expert to communicate his knowledge, in the form of facts and rules, to the expert system. Above all, this specialist interface must be fast, as well as being convenient to use, easy to check and simple to test. If the input system is not acceptable to the specialist, the expert system will not succeed.

When it is possible, a human intermediary, trained in the specialty of the subject, as well as being fully conversant with the software system, is the most effective means of transferring the expert's knowledge into the system. This has been the approach which we have found in practice to be most successful, but it needs to be stressed that the intermediary must be highly trained in the specialty. A computer scientist will not normally be able to do this. For one successful information system in physics, we have used an intermediary who, herself, had a PhD in physics. The intermediary could then converse easily and rapidly with the specialists in the subject and the specialist's time was kept to an absolute minimum[6].

12.12   EXPERT SYSTEMS

To judge by the number of books and published papers on expert systems, there can be no doubt that their development in recent years has started a major new field of application for computers, which is also quite different from traditional computing. However, the number of actual working systems on which we can obtain public information is quite limited. In the early days, up to about 1980, most systems were developed at universities, notably Stanford, and reported in some detail in the literature. However, as the potential for problem solving in industry of these new systems has become apparent, industry has been taking the lead through its own research and through its support of research at universities. These industrial expert systems usually model expertise built up over many years on some part of a company's activities and they can have large commercial value. So considerable secrecy surrounds the methods used in great bulk of these industrial systems. It is known that they are being used widely, more than experimentally, in such areas as oil exploration and in fault diagnosis of complex computer and other systems. It is also known that all of the successful systems are highly specialized, covering very limited areas.

The most commonly described system in text books and articles is one of the earliest medical expert systems, called MYCIN, developed at Stanford, which specializes in infections of the lower abdomen[7]. Several expert systems today are modelled on this early system. However, since our space is limited, we are not able to discuss the MYCIN system nor its approximate methods of dealing with 'certainty' rather than probability. (In any case MYCIN is described in detail in almost every other text, see [1–4]). Since this book is aimed at scientists and engineers, we give a brief description of the mineral exploration system, called Prospector[8] and for this reason the discussion on probability already given in

Sections 12.5 and 12.6 was based on Prospector rather than on the approximate methods adopted in MYCIN (though the MYCIN approach is more common).

Prospector is an expert system to aid field geologists to give advice on the most likely sites for the existence of a particular mineral deposit. Separate versions exist for each mineral: one for zinc, one for uranium and so on. Knowledge is stored in the form of production rules similar to the ones we have been discussing, e.g. a simplified rule might be:

If there are abundant quartz sulphide veinlets then potassium zinc is likely, confidence 0.7.

A network is built up of 'spaces', equivalent to our facts, linked together in a network by rules. Initial probabilities are given to the spaces and through a series of questions to the field Geologist, aimed at some hypothesis, the probabilities of each of the basic spaces is determined. By chaining, the probabilities of other relevant spaces are changed in order. A typical question might be

What is the age of the greenstone belt?

Archaean

What is your confidence in this estimate?

4

Confidence is measured on a scale from 0 to 5, 5 meaning certainty. Bayesian theory is used to compute the probabilities of each consequence of each rule affected by an answer as in Section 12.6.

As we have stated, several versions of Prospector have been developed for different minerals. One version shows maps, shading the areas to indicate the most likely areas to find the particular mineral.

It is interesting that Prospector itself is not in every day use. However, it has pointed the way for several later commercial systems, and geological exploration is now one of the major applications of expert systems, particularly for oil explorations which includes analysis of seismic data. However, as we have stated, there is little detailed public knowledge on the methods used in these systems.

12.13   EXPERT SYSTEM SHELLS

An expert system like Prospector or MYCIN consists of facts and rules and a software system, called the inference engine, for computing the solution to problems. In these early systems the rules and software are integrated into a single system, with numerous special cases, exceptions and modifications built into the system to make it successful. If the rules and facts and other special features which are relevant to geology or medicine or other specialty, are removed from the system, we are left with a shell system, with no data left within the system. In the case of the

MYCIN Expert System, after the knowledge base was removed, the shell was called EMYCIN (standing for Empty MYCIN).

A shell system, like EMYCIN, is purely a software system and to have any function it needs knowledge in the form of facts and rules to be added to the system; so it is equivalent to a database management system such as dBASE III or Oracle. A large number of such shell systems are now on the market, both for small and for large computers. One of the common systems is Xi by Expertech, another is Expertise by J. Perrone and Associates. Rather than attempt to review these systems here, we refer the reader to frequent reviews of expert systems in the popular computer literature, such as *Byte*.

Generally these shell systems are quite easy to use, just as modern database management systems are easy to use. However, to model the expertise of a human specialist in a complex domain it is frequently necessary to add a substantial amount of special software to deal with special cases and to include unusual rules which do not take a conventional form. This is particularly important for expert systems in science and engineering as you may need to include calculations within the expert system and this will need additional software. Even in the case of the original system MYCIN, when the facts and rules of MYCIN were added to its own shell EMYCIN it was found not to perform well; it was not as good as the original system, MYCIN. Special software was needed for 'fine tuning'. It is therefore important, in the selection of an expert system shell, to select one with a good interface to a high level programming language.

12.14 SUCCESSFUL SMALL EXPERT SYSTEMS

Expert system shells can be used with great success on a wide range of problems and are particularly successful when used on PCs to solve simple straightforward problems. A common example, is an expert system to aid first line maintenance on some complex item of equipment. A very simple example, a system to diagnose faults in an inexpensive printer, has already been given in Section 12.2. In a more realistic example there may be 100 or more faults which might affect the equipment, but 20 of them may be common and relatively easy to diagnose and correct.

This is the kind of situation which is particularly suitable for an expert system shell; it should aim to diagnose these 20 most frequent faults and print or display the remedies which enable the junior engineer or technician to correct most faults which occur. The senior engineer, the expert, need only be called when the fault is one of the rare and less easily diagnosed faults. This saves the time of the expert, saves costs and reduces the average time taken to correct faults since the time of the senior engineer is no longer a bottle-neck restricting the user service. In this type of maintenance situation (and in many other simple expert systems) it is not worth spending a great deal of time and effort building a more complex expert system to model more fully all of the knowledge of the senior engineer. The cost is high and the return small, particularly if the equipment to be maintained is soon likely to be replaced by a new model and a new expert system for fault finding will be required.

12.15   PROGRAMMING LANGUAGES

If you wish to build an expert system yourself, without using a shell system, we first ask you to think again, as there is little point wasting time reinventing wheels and redoing what can be purchased at a low cost. A decision to build your own system may be justified if your expert system is going to be used very frequently by large numbers of users. Examples are found in areas such as financial services; one of us (FJS) is a director of a company (Expert Information Systems) providing such expert systems which can be accessed by over 12 000 customers for advice using a data network. In this circumstance, efficiency is paramount and although a shell was used to model part of the expert system when the company was founded, the software was all rewritten in C which is a fast and flexible high level language that uses the resource of the time shared computer system efficiently. If a shell was still used, the number of customers able to access the system at one time would be down by a large factor and the company would not be in business.

Unlike financial services, in science or engineering it is likely that the number of users of an expert system will be small and the frequency of access to the system will be low. In this case it is likely to be implemented on a large PC, or on a mini-computer such as a MicroVAX, and the cost of running the system will not be important. A shell system is then appropriate. If the shell system has a good interface to a high level programming language you can then add software to the shell to widen its capabilities.

A decision to build your own system may still be justified if you find that the expert system you need requires a great deal of software over and above the kind of inference machine provided by proprietary expert systems. These normally can only process logical rules, but often in science or engineering a great deal of calculations are needed and logical programming is a minor part of the processing needed.

You could write your expert system in any language: FORTRAN, PASCAL, C, or even BASIC[1], and expert systems have been written successfully in all of these procedural languages. Your choice of language may be dictated by the other software in your total system. However, if you can choose any language you will find it much easier to write the inference engine of an expert system in a declarative language like LISP, or PROLOG[9]. Compilers are widely available for both languages, but you will find PROLOG easier to use. It can be learned fairly easily from a good text[10].

12.16   KNOWLEDGE BASED MANAGEMENT SYSTEMS

It is not uncommon for a combined database and expert system to be needed[11].

For example, for computer aided design an engineering company may already have a database of drawings of items designed by the company along with many modifications; it may also have a database of components and materials. However, the experience of an expert designer may be crucial in the selection of an old design as the basis of a new one or in the selection of materials or components; so a database system is today not sufficient for a competitive engineering design team.

One possibility is to combine an expert system with a database management system by passing data from one to the other, and by executing one then the other. However, this process is slow and since moving from one system to the other may be very frequent during the interrogation of the whole system by a user, the process is slow.

The alternative is either to build suitable database structures onto an expert system shell or to add some reasoning capabilities to the user interface of a traditional relational or other database management system. This second approach we think is the more promising of the two and it has been demonstrated that it is not difficult to build a PROLOG written reasoning capability on top of a database management system such as Oracle. However, for some problems the first approach, a modified expert system, will be more appropriate.

Such combined systems are still new and the subject of research[14, 15]. So it is still early to give advice on them. When an expert system is extended to widen the types of data and processing possible, it is commonly called an Intelligent Knowledge Based System (IKBS) and the software system (or shell) is called a Knowledge Based Management System (KBMS)[16]. But all such systems are still experimental and are still the subject of debate by specialists in computer science or in artificial intelligence.

12.17   OBJECT-ORIENTED KNOWLEDGE BASE SYSTEMS

Object-oriented knowledge based systems are also the subject of a great deal of research and discussion at present and we believe that they may enable significant advances in knowledge processing systems within the next decade. We discuss these briefly here to allow the reader to begin to follow the literature more easily. We have earlier described object oriented programming and object-oriented database systems in Section 9.6.

A traditional expert system stores facts and logical rules only. This permits the solution of problems by the processing of these logical rules by an inference engine. But, as we have made clear, this is not sufficient in many practical expert systems, unless they are unusually simple ones. Most larger expert systems need special software written into the system to deal with special cases and to model processes which cannot be modelled by our simple logical rules. So the storage of facts and rules alone is not sufficient in practice and each expert system has substantial special **implicit** code built into the inference engine to model additional processes and without which it would not be effective. So every expert system is, in a sense, a once-off system.

A more general system would permit the **explicit** storage of the processes in the knowledge base in a more flexible form, i.e. as code in the form of modules, or routines or whole programs. To do this we need a system which can store and process a richer variety of objects than, for example, a relational database system can handle. Object-oriented database systems (OODS) mentioned in Section 9.6 are ideal structures for this purpose.

In an OODS the objects, which are data models of entities in the real world, may include data and the processes operating on the data. These processes, represented by explicit code stored with the data, may include the equivalent of the logical rules found in a traditional expert system, but in addition, because any code, any module, any program can be included as part of an object (or inherited by an object) the range of processes which can be implemented is as wide as the knowledge base permits. For example, objects may include processes which include the relevant laws of physics and chemistry, search through large knowledge bases, carry out substantial numeric processing, process images, retrieve information from text, analyze language and so on. So the object-oriented approach is a suitable wide vehicle for the storage and implementation of a very flexible knowledge base system. In this approach, complex objects, including processes and relationships between objects, replace the facts and rules of an expert system. Within a decade such systems may be common-place tools in the armory of the knowledge engineer.

REFERENCES

[1] C. Naylor, *Build your own Expert System*, Sigma Technical Press, Willmslow, Cheshire, UK, p249, 1983.

[2] J. L. Alty and M. J. Coombs, *Expert Systems: Concepts and Examples*, NCC Publications, Manchester, England, p209, 1984.

[3] W. J. Black, *Intelligent Knowledge Based Systems: an Introduction*, Van Nostrand Reinhold (UK), Wokingham, England, p159, 1986.

[4] A. Bundy, An Expert System for Medical Diagnosis, in *Intelligence Knowledge Based Systems*, Eds. T. O'Shea, J. Self and G. Thomas, Harper and Rowe, London, 36–51, 1987.

[5] R. A. Frost, *Introduction to Knowledge Base Systems*, Collins Professional and Technical Books, London, p677, 1986.

[6] F. J. Smith and J. G. Hughes, An Expert Driven Database System, *CODATA Conference Proceedings*, Karlsruhe, Germany, in press 1989.

[7] D. D. Wolfgram, T. J. Dear and C. S. Galbraith, *Expert Systems for the Technical Professional*, J. Willey & Sons, N.Y., p289, 1988.

[8] E. Shortliffe, *Computer based Medical Consultation: MYCIN*, Elsevier Publication Inc., New York, 1976.

[9] R. Dudda, J. Gaschnig and P. Hart, Model Design in the Prospector Consultant System for Mineral Exploration, in *Expert Systems in A Microelectronic Age*, Ed. D. Michie, Edinburgh University Press, Edinburgh, 153–167, 1979.

[10] P. Harman, *Analyzing Expert System Building Tools*, Harman Associates, 1987.

[11] A. Sloman, Artificial Intelligence Languages, in *Intelligent Knowledge-based Systems*, Eds. T. O'Shea, J. Self and G. Thomas, Harper and Rowe, London, 15–35, 1987.

[12] D. Crookes, *Programming in Prolog*, Academic Press, London, 1987.

[13] G. Wiederhold, Knowledge Versus Data, in *Knowledge Based Management Systems*, Eds. M. L. Brodie and J. Mylopoulos, Springer Verlag, New York, 77–82, 1986.

[14] U. Dayal and J. M. Smith, PROBE: A Knowledge-oriented Database Management System, in *Knowledge Base Management Systems*, Eds. M. L. Brodie and J. Mylopoulos, Springer-Verlag, New York, 227–257, 1986.

[15] P. M. D. Gray, I. G. Archibald and K. Lunn, Interfacing a Knowledge-based System to a Large Database, *Knowledge Eng. Rev.*, **4**, 31–51, 1989.

[16] J. Mylopoulos, On Knowledge Base Management Systems, in *Knowledge Base Management Systems*, Eds. M. L. Brodie and J. Mylopoulos, Springer-Verlag, New York, 3–8, 1986.

Index